Probability and Mathematical Statistics (Continued)

RAO • Linear Statistical Inference and Its Applications, *Second E*[illegible]
ROBERTSON, WRIGHT, and DYKSTRA • [illegible] e
ROGERS and WILLIAMS • Diffusions, Ma[illegible] me II: Îto Calculus
ROHATGI • A Introduction to Probability T[illegible]
ROSS • Stochastic Processes
RUBINSTEIN • Simulation and the Monte C[illegible]
RUZSA and SZEKELY • Algebraic Probabil[illegible]
SCHEFFE • The Analysis of Variance
SEBER • Linear Regression Analysis
SEBER • Multivariate Observations
SEBER and WILD • Nonlinear Regression
SERFLING • Approximation Theorems of Mathematical Statistics
SHORACK and WELLNER • Empirical Processes with Applications to Statistics
STAUDTE and SHEATHER • Robust Estimation and Testing
STOYANOV • Counterexamples in Probability
STYAN • The Collected Papers of T. W. Anderson: 1943–1985
WHITTAKER • Graphical Models in Applied Multivariate Statistics
YANG • The Construction Theory of Denumerable Markov Processes

Applied Probability and Statistics

ABRAHAM and LEDOLTER • Statistical Methods for Forecasting
AGRESTI • Analysis of Ordinal Categorical Data
AGRESTI • Categorical Data Analysis
ANDERSON and LOYNES • The Teaching of Practical Statistics
ANDERSON, AUQUIER, HAUCK, OAKES, VANDAELE, and WEISBERG • Statistical Methods for Comparative Studies
ASMUSSEN • Applied Probability and Queues
*BAILEY • The Elements of Stochastic Processes with Applications to the Natural Sciences
BARNETT • Interpreting Multivariate Data
BARNETT and LEWIS • Outliers in Statistical Data, *Second Edition*
BARTHOLOMEW, FORBES, and McLEAN • Statistical Techniques for Manpower Planning, *Second Edition*
BATES and WATTS • Nonlinear Regression Analysis and Its Applications
BELSLEY • Conditioning Diagnostics: Collinearity and Weak Data in Regression
BELSLEY, KUH, and WELSCH • Regression Diagnostics: Identifying Influential Data and Sources of Collinearity
BHAT • Elements of Applied Stochastic Processes, *Second Edition*
BHATTACHARYA and WAYMIRE • Stochastic Processes with Applications
BIEMER, GROVES, LYBERG, MATHIOWETZ, and SUDMAN • Measurement Errors in Surveys
BIRKES and DODGE • Alternative Methods of Regression
BLOOMFIELD • Fourier Analysis of Time Series: An Introduction
BOLLEN • Structural Equations with Latent Variables
BOX • R. A. Fisher, the Life of a Scientist
BOX and DRAPER • Empirical Model-Building and Response Surfaces
BOX and DRAPER • Evolutionary Operation: A Statistical Method for Process Improvement
BOX, HUNTER, and HUNTER • Statistics for Experimenters: An Introduction to Design, Data Analysis, and Model Building
BROWN and HOLLANDER • Statistics: A Biomedical Introduction
BUCKLEW • Large Deviation Techniques in Decision, Simulation, and Estimation
BUNKE and BUNKE • Nonlinear Regression, Functional Relations and Robust Methods: Statistical Methods of Model Building

Continued on back end papers

*Now available in a lower priced paperback edition in the Wiley Classics Library.

Alternative Methods of Regression

Alternative Methods of Regression

DAVID BIRKES
Oregon State University

YADOLAH DODGE
University of Neuchâtel

A Wiley-Interscience Publication
JOHN WILEY & SONS, INC.
New York • Chichester • Brisbane • Toronto • Singapore

To my parents.
—Dave

To Kay.
—Yadolah

This text is printed on acid-free paper.

Library of Congress Cataloging in Publication Data:

Birkes, David
Alternative methods of regression / David Birkes, Yadolah Dodge.
p. cm.
"A Wiley-Interscience publication."
Includes index.
ISBN 0-471-56881-3
1. Regression analysis. I. Dodge, Yadolah, 1944– II. Title.
QA278.2.D64 1993
519.5′36—dc20 92-31165

Printed in the United States of America

10 9 8 7 6 5 4 3 2 1

Contents

Preface

Since its discovery almost 200 years ago, least squares has been the most popular method of regression analysis. A statistics book with the word "regression" in its title, without any qualifying adjective such as "robust" or "nonparametric" or "alternative", can be assumed to be about least-squares regression. Over the last two or three decades, however, there has been increasing interest in other methods. This is due partly to discoveries of deficiencies in the least-squares method and partly to advances in computer technology, which have made the computational complexity of other methods a relatively unimportant consideration. Numerous research articles have now been published on alternative approaches to regression analysis.

Development of these approaches continues and it is likely that further research and experience will lead to modifications and improvements. But enough knowledge and experience have already been accumulated to be able to say that currently proposed alternative methods give reasonable results, have worthwhile advantages over least-squares methods, and can be recommended for practical use.

This book provides an introduction to a variety of regression methods. Chapters 1 and 2 deal with the general idea of linear regression without specifying any particular method. Chapter 3 is a review of least-squares regression and serves as a reference for comparison with other methods. Chapters 4 through 8 present five alternative methods: least-absolute-deviations, robust M-, nonparametric rank-based, Bayesian, and ridge regression. We do not intend to imply that these five methods are the best ones available (which in any case depends on one's definition of "best"). We have chosen them because they represent a range of approaches to regression analysis and they have received a fair amount of attention in the statistical literature. A number of other methods are mentioned briefly in Chapter 10.

Chapters 4 through 8 can be read in any order after reviewing the first three chapters. The formats of these chapters are similar. Each chapter (except Chapter 8, which does not discuss testing) has sections on estimating the regression line for simple regression data, testing the slope of the regression line, estimating the regression coefficients for multiple regression data, and testing the regression coefficients. In each section a formula is displayed or an algorithm is described for calculating the estimate or test, and justification is given for why the procedure makes sense. Each chapter is intended to be a relatively quick introduction to the method, and so mathematical derivations, bibliographical citations, and other details are deferred to notes at the end of the chapter. If the reader is unclear about some point in the text, clarification can sometimes be found in the notes. (The notes are numbered according to the section they pertain to.) Chapter 9 contains more information about the six methods and compares them according to their theoretical properties and their performances on some real data sets.

The descriptions of the procedures are sufficient to enable a reader with computer programming ability to write programs to implement them. Worked examples are included. (Most numbers are displayed to four significant digits, but calculations were carried out using greater accuracy.) The section on computation near the end of each chapter mentions existing programs and packages that perform these procedures.

The ideal prerequisites for this book are a course in least-squares regression, a course in mathematical statistics, and familiarity with matrix notation. In order to make the book as readable as possible for someone whose knowledge in these prerequisite areas is rusty, a review of least-squares regression and some basic statistical concepts and definitions are presented in Chapter 3.

It is hoped that readers of this book will learn enough about how and why alternative methods work that they will be motivated to apply them.

We would like to thank the following people for reading parts of the manuscript and providing helpful comments: Gilbert Bassett, Thomas Hettmansperger, Peter Huber, Jana Jurečková, Roger Koenker, Peter Rousseeuw, Adrian Smith, Stephen Stigler, and especially Dennis Cook. We would also like to thank Séverine Pfaff for her expert typing. One author (DB) is grateful to the University of Neuchâtel and Oregon State University for support while on leave during the academic year 1990–91.

DAVID BIRKES
Corvallis, Oregon

YADOLAH DODGE
Neuchâtel, Switzerland

Acknowledgments

We thank the authors and publishers of tables and data for permission to use the following materials:

- Table 1.1, Draper, N. and Smith H., *Applied Regression Analysis* (1981, p. 406), ©Reprinted by permission of John Wiley & Sons, New York.
- Table 1.3, ©(1986) World Resources Institute publications, Washington D.C. and ©(1986) Basic Book, New York.
- Table 1.4, Galton (1986), by permission of The Royal Anthropological Institute, London.
- Table 3.1, Daniel, C. and Wood, F. S., *Fitting Equations to Data* (1980, p. 46), ©Reprinted by permission of John Wiley & Sons, New York.
- Table 4.1, Andrews, D. F. and Herzberg, A. M., *Data* (1985, p. 409), ©Springer-Verlag, New York.
- Table 5.1, Reprinted by permission from *Statistics : the Exploration and Analysis of Data*, Devore, J. and Peck, R. (1986, pp. 471–472), ©1986 by West Publishing Company. All right reserved.
- Table 5.3, Reprinted by permission of ©SAS Institute Inc., *SAS/STAT* (*R*) *User's Guide*, *Version* 6, Fourth Edition, Volume 2, Cary, North Carolina: SAS Institute Inc., 1990, 1686 pp.
- Table 6.1, Reprinted from Gunst, R. F. and Mason, *Regression Analysis and Its Applications*, (1980, p. 367). Courtesy of ©Marcel Dekker, Inc. New York.
- Table 6.3, Chattergee, S. and Price, B., *Regression Analysis by Example*. (1977, p. 99). Reprinted by permission of ©John Wiley & Sons, New York.
- Table 7.2, Daniel, C. and Wood, F. S., *Fitting Equations to Data*. (1980, p. 110). Reprinted by permission of ©John Wiley & Sons, New York.
- Table 8.1, S. Weisberg, *Applied Linear Regression*, (1985, p. 57). Reprinted by permission of ©John Wiley & Sons, New York.
- Table 8.1, Reprinted by permission of the American Chemical Society, Washington D.C. and of A. Hald, *Statistical Theory with Engineering Applications*, (1952, p. 647), John Wiley & Sons, New York with permission from the author.
- Table 9.2, Brownlee, K. A., *Statistical Theory and Methodology in Science and*

Engineering, (1965, p. 454). Reprinted by permission of ©John Wiley & Sons, New York.

- Tables of Student and F distributions, Reproduced with permission of the Biometrika Trustees from E. S. Pearson and H. O. Hartely (1966), *Biometrika Tables for Statisticians*, Vol. 1, 3rd ed., London: Cambridge University.

CHAPTER 1

Linear Regression Analysis

Ho, saki, haste, the beaker bring,
Fill up, and pass it round the ring;
Love seemed at first an easy thing—
But ah! the hard awakening.

HAFEZ SHIRAZI: *Persian Poet (1348–1398)*

1.1 INTRODUCTION

Statistics is the science of making sense out of data. For various types of data, various statistical techniques have been developed for extracting relevant information from the data—for determining what the data "mean". One common type of data occurs when the values of several variables are measured for each of several units. For example, in a medical study, the age, weight, height, and blood pressure of a group of 100 subjects might be recorded. Here the values of four variables are measured for each of 100 units. Or, in an industrial experiment, the reaction time, reaction temperature, and product yield of 20 batches of a chemical might be recorded. Here the values of three variables are measured for each of 20 units. Often such data are collected for the purpose of seeing how one of the variables depends on the others. In the medical study one would be interested in how a subject's blood pressure depends on his or her age, weight, and height. In the industrial experiment one would be interested in how the product yield of a batch of chemical depends on the reaction time and reaction temperature of the batch. Linear regression analysis is a commonly used statistical technique for dealing with such data. The first two chapters introduce the general concept of linear regression analysis and Chapters 3 through 10 present a variety of particular methods for implementing it.

Table 1.1 Turnip Green Data

Identification Number of Plant	Vitamin B_2 (Y)	Sunlight (X_1)	Soil Moisture (X_2)	Air Temperature (X_3)
1	110.4	176	7.0	78
2	102.8	155	7.0	89
3	101.0	273	7.0	89
4	108.4	273	7.0	72
5	100.7	256	7.0	84
6	100.3	280	7.0	87
7	102.0	280	7.0	74
8	93.7	184	7.0	87
9	98.9	216	7.0	88
10	96.6	198	2.0	76
11	99.4	59	2.0	65
12	96.2	80	2.0	67
13	99.0	80	2.0	62
14	88.4	105	2.0	70
15	75.3	180	2.0	73
16	92.0	180	2.0	65
17	82.4	177	2.0	76
18	77.1	230	2.0	82
19	74.0	203	47.4	76
20	65.7	191	47.4	83
21	56.8	191	47.4	82
22	62.1	191	47.4	69
23	61.0	76	47.4	74
24	53.2	213	47.4	76
25	59.4	213	47.4	69
26	58.7	151	47.4	75
27	58.0	205	47.4	76

Source: Draper and Smith (1981, p. 406).

1.2 EXAMPLE

The data in Table 1.1 were collected for a study of vitamin B_2 in turnip greens. Three factors thought to influence the amount of vitamin B_2 in a plant are sunlight, soil moisture, and air temperature. On 27 different occasions the investigators picked the leaves of a turnip plant and measured the concentration of vitamin B_2 (in milligrams per gram of turnip greens), the amount of radiation during the preceding half-day of sunlight (in relative gram calories per square centimeter per minute), the average soil moisture tension, and the air temperature (in degrees Fahrenheit). From these data we want to determine how the amount of vitamin B_2 in the leaves of a turnip plant depends on the factors of sunlight, soil moisture, and air temperature.

1.3 THE LINEAR REGRESSION MODEL

Notation and Terminology. It is convenient to use the following notation:

$$y_i = \text{concentration of vitamin B}_2 \text{ for the } i\text{th plant}$$
$$x_{i1} = \text{sunlight measurement for the } i\text{th plant}$$
$$x_{i2} = \text{soil moisture measurement for the } i\text{th plant}$$
$$x_{i3} = \text{air temperature for the } i\text{th plant}$$

When referring to a particular plant for which data have been collected, we use the notation $y_i, x_{i1}, x_{i2}, x_{i3}$. When talking in general about an unspecified plant, we use the notation Y, X_1, X_2, X_3.

The variable Y, which is thought to depend on the other variables, is called the *dependent variable* or *response variable*. The other variables, X_1, X_2, and X_3, are called *independent variables* or *explanatory variables*.

Formulation of a Model for the Example. To analyze the data in the example, we begin with the plausible postulate that the concentration of vitamin B_2 in the leaves of a turnip plant (Y) is, approximately, a function of sunlight (X_1), soil moisture (X_2), and air temperature (X_3). We do not expect to be able to express Y as an exact mathematical function of X_1, X_2, and X_3 because, as is generally true of real-life processes, the production of vitamin B_2 is more complicated than that. But it seems reasonable to suppose that part of the concentration of vitamin B_2 in a leaf can be explained by these three explanatory variables. Let us write $f(X_1, X_2, X_3)$ to express the part of Y explained by the X's. A useful model is obtained by supposing that Y differs from $f(X_1, X_2, X_3)$ by a random quantity e. (We cannot expect this model, or any model, to completely describe reality, but it is more realistic than assuming $Y = f(X_1, X_2, X_3)$ exactly.) The equation $Y = f(X_1, X_2, X_3) + e$ is a *regression model* in which f is the *regression function* and e is a *random error*.

A *linear regression model* is a regression model in which the regression function is linear. A linear regression model for the example is

$$Y = \beta_0 + \beta_1 X_1 + \beta_2 X_2 + \beta_3 X_3 + e.$$

The form of the function is specified to be linear but the coefficients β_0, β_1, β_2, and β_3 are not specified; they are unknown parameters. They are called *regression coefficients* or *regression parameters*. The error e is assumed to be a random variable with mean 0.

The reason for the word "linear" in "linear regression" is of course the linear form of the regression function. The reason for the word "regression" is explained in Section 1.7.

Table 1.2 Array of Data Consisting of n Observations of a Response Variable and p Explanatory Variables

Observation Number	Y	X_1	X_2	$\cdots$	X_p
1	y_1	x_{11}	x_{12}	$\cdots$	x_{1p}
2	y_2	x_{21}	x_{22}	$\cdots$	x_{2p}
$\vdots$	$\vdots$	$\vdots$	$\vdots$		$\vdots$
n	y_n	x_{n1}	x_{n2}	$\cdots$	x_{np}

Linearity. The essential feature of a linear regression function is that it is a sum of terms $\beta_j X_j$, where β_j is an unknown parameter and X_j is a known value obtained from the data. The manner in which X_j is obtained from the data is not important. For example, suppose we have data on the ages (Y) and diameters (X) of pine trees. One linear regression model for these data is $Y = \beta_0 + \beta_1 X + e$. Another linear regression model we could try is $Y = \beta_0 + \beta_1 X + \beta_2 X^2 + e$. The latter model might look like a quadratic model rather than a linear model because it involves X^2. But if we simply change notation by setting $X_1 = X$ and $X_2 = X^2$, then we can write the model as $Y = \beta_0 + \beta_1 X_1 + \beta_2 X_2 + e$, which falls under the definition of a linear regression model.

The General Model. In general, suppose we have data consisting of n observations of a response variable Y and p explanatory variables $X_1, X_2, \ldots, X_p$. It is convenient to arrange the data in an array as in Table 1.2. Such data are sometimes called "regression data" because it is natural to think of using a regression model to analyze them. The *linear regression model* is the equation

$$Y = \beta_0 + \beta_1 X_1 + \beta_2 X_2 + \cdots + \beta_p X_p + e \tag{1.1}$$

In terms of the observed data, the model is

$$y_i = \beta_0 + \beta_1 x_{i1} + \beta_2 x_{i2} + \cdots + \beta_p x_{ip} + e_i \tag{1.2}$$

for $i = 1, 2, \ldots, n$. Also included in the model is the assumption that the random errors $e_1, e_2, \ldots, e_n$ can be regarded as a random sample of independent draws from a single population having a mean of 0.

When Is the Model Appropriate? If we have a set of data which we want to analyze by linear regression, how can we tell if the linear regression model is appropriate for these data? In the first place, it must be realized that a theoretical model can never *exactly* describe a real-world process such as the production of vitamin B_2 in turnip leaves, because the real world is too complex. But for the model to be useful, the data should appear to *approximately* follow the model. The linear regression model makes several assumptions about the data, including linearity of the function of the explanatory variables, independence of the random errors, and equality of the variances of the random errors. Procedures have been developed for checking the approximate validity of these assumptions. Some of these procedures are discussed in Chapter 2.

1.4 ESTIMATING THE REGRESSION COEFFICIENTS

Returning to the turnip green data in Table 1.1, let us suppose (at least tentatively, subject to reconsideration) that the linear regression model in Section 1.3 is appropriate for these data. That is, we suppose $Y = \beta_0 + \beta_1 X_1 + \beta_2 X_2 + \beta_3 X_3 + e$. The regression coefficients β_0, β_1, β_2, and β_3 are unknown parameters which need to be estimated. The basic idea of the model is that the linear function $\beta_0 + \beta_1 X_1 + \beta_2 X_2 + \beta_3 X_3$ is "close" to the value of Y. So, in using the observed data to estimate the regression coefficients, the basic idea is to estimate them by values $\hat{\beta}_0$, $\hat{\beta}_1$, $\hat{\beta}_2$, and $\hat{\beta}_3$ such that the *fitted* y-value,

$$\hat{y}_i = \hat{\beta}_0 + \hat{\beta}_1 x_{i1} + \hat{\beta}_2 x_{i2} + \hat{\beta}_3 x_{i3}$$

is "close" to the observed y-value, y_i, for $i = 1, 2, \ldots, 27$. How exactly do we measure the collective closeness of the fitted y-values to the observed y-values? The various regression methods presented in the following chapters correspond to different ways of measuring closeness.

Suppose we have chosen a regression method and have obtained estimates $\hat{\beta}_0$, $\hat{\beta}_1$, $\hat{\beta}_2$, and $\hat{\beta}_3$. The equation $\hat{Y} = \hat{\beta}_0 + \hat{\beta}_1 X_1 + \hat{\beta}_2 X_2 + \hat{\beta}_3 X_3$ is called the estimated regression equation or the *fitted* regression equation. This equation has several uses. First, it serves as a concise description of a major pattern in the observed data. It describes an approximate relationship between vitamin B_2 and the other three variables. The values y_i tend to be close to $\hat{\beta}_0 + \hat{\beta}_1 x_{i1} + \hat{\beta}_2 x_{i2} + \hat{\beta}_3 x_{i3}$. Second, it could be used to predict the concentration of vitamin B_2 in future harvests of turnip greens with possibly different degrees of sunlight, soil moisture, and air temperature. If in the future we measure the values X_1, X_2, and X_3, we could predict that the value of Y will be near $\hat{Y}$. Third, the fitted regression equation provides a basis for testing the significance of the relationship between the response variable and the explanatory variables.

1.5 TESTING THE SIGNIFICANCE OF THE RELATIONSHIP

A good model balances two competing requirements. On the one hand, the model should be complex enough to approximate the real-world phenomenon it describes. On the other hand, the model should be as simple as possible, because the simpler it is, the more comprehensible it is. Therefore, if we have two models that give approximately the same degree of agreement with reality, we should prefer the simpler model.

In the context of linear regression analysis, this principle of simplicity is applied in the following way. From the full linear regression model $Y = \beta_0 + \beta_1 X_1 + \cdots + \beta_p X_p + e$ we drop out one or more of the explanatory variables, say $X_{q+1}, \ldots, X_p$, to obtain a reduced model $Y = \beta_0 + \beta_1 X_1 + \cdots + \beta_q X_q + e$. We then perform a test to see if there is a significant difference between the full model and the reduced model. If there is no significant difference, then the two models are approximately equivalent and so we prefer the simpler one, the reduced model.

For example, to analyze the turnip green data in Table 1.1, we could compare the full model $Y = \beta_0 + \beta_1 X_1 + \beta_2 X_2 + \beta_3 X_3 + e$ with the model $Y = \beta_0 + e$ to see whether vitamin B_2 has a significant dependence on any of the three explanatory variables. Of course we expect such a dependence but we should check whether the data confirm our expectations. We could also compare the full model with the model $Y = \beta_0 + \beta_1 X_1 + \beta_2 X_2 + e$. This would test whether the dependence of vitamin B_2 on air temperature is still significant after its dependence on sunlight and soil moisture is taken into account. And several other similar tests could be performed.

The different regression methods presented in the following chapters entail different procedures for performing such tests.

1.6 THE NEED FOR ALTERNATIVE METHODS

When one wants to apply the linear regression model to a set of data, there are various methods one could use to estimate the regression coefficients. The most popular one is called the method of least squares, which is presented in Chapter 3. This method, however, has weaknesses and consequently sometimes statisticians prefer to use alternative regression methods.

A major weakness of the method of least squares is illustrated by the following example. In Table 1.3 are listed the 14 countries in North and Central America with populations over one million people in 1985. For each country the table shows its birth rate (the number of births per year per 1000 people) for 1980–1985 and its urban percentage (the percentage of the population living in cities of over 100,000) in 1980.

Table 1.3 Birth Rate Data

Country	Birth Rate (Y)	Urban Percentage (X)
Canada	16.2	55.0
Costa Rica	30.5	27.3
Cuba	16.9	33.3
Dominican Republic	33.1	37.1
El Salvador	40.2	11.5
Guatemala	38.4	14.2
Haiti	41.3	13.9
Honduras	43.9	19.0
Jamaica	28.3	33.1
Mexico	33.9	43.2
Nicaragua	44.2	28.5
Panama	28.0	37.7
Trinidad/Tobago	24.6	6.8
United States	16.0	56.5

Source: World Resources Institute (1986).

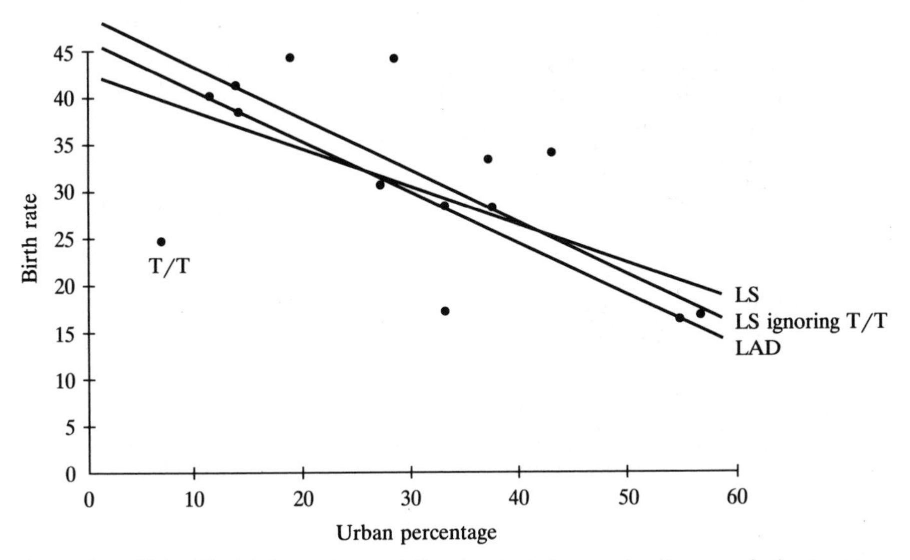

Figure 1.1 Plot of the birth rate data with three estimated regression lines, one by least squares (LS) using all the data, one by LS ignoring Trinidad/Tobago (T/T), and one by least absolute deviations (LAD).

To investigate the relationship between these two variables we can plot the data as in Figure 1.1. The points in the plot show, at least roughly, a pattern sloping downward from left to right. That is, there seems to be a tendency for countries with higher percentages of urban dwellers to have lower birth rates. To describe this tendency quantitatively, we can fit a linear regression equation to the data. If we use the method of least squares, the equation is $\hat{Y} = 43 - 0.40X$.

Note how the point representing Trinidad/Tobago stands apart from the rest of the points. A weakness of the least-squares method is that it allows such "outlying" data points to have too much influence. If we eliminate Trinidad/Tobago from the analysis and fit a least-squares linear regression equation to the remaining 13 points, we obtain $\hat{Y} = 49 - 0.55X$, which is noticeably different from the first equation. It seems unwise to allow a single country to have such a large influence on our analysis.

Alternative regression methods are available which restrain the influence of outlying data points. For example, we can use the method of least absolute deviations, which is presented in Chapter 4. If we fit a linear regression equation to the same data by the method of least absolute deviations, the equation is $\hat{Y} = 46 - 0.54X$. See Figure 1.1. Without Trinidad/Tobago, the equation is exactly the same.

M-regression, presented in Chapter 5, and nonparametric regression, presented in Chapter 6, are also resistant to the influence of outliers. The least-squares method of regression performs best if the population of errors has a normal distribution. If it is thought that the distribution of errors may not be normal (which is sometimes indicated by the occurrence of outliers), then least-squares estimates and tests may be much less efficient than those provided by the methods of least absolute deviations, M-regression, or nonparametric regression.

Even if the distribution of errors is normal, the accuracy of least-squares estimates can sometimes be improved by using the ridge estimates described in Chapter 8. When one has previous knowledge about the type of data one is analyzing, it is possible to incorporate this knowledge into the analysis by using the Bayesian methods presented in Chapter 7. Further discussion of the advantages and disadvantages of least-squares regression and the various alternative methods is given in Chapter 9.

1.7 THE ORIGIN OF THE WORD "REGRESSION"

Francis Galton introduced the concept of regression in a paper on "Typical Laws of Heredity" presented in England in 1877. In a study of inheritance in sweet peas, he discovered an interesting relationship between the diameters of parent sweet peas and the diameters of their offspring. Table 1.4 and

Table 1.4 Galton's Sweet Pea Data

Diameter of Parent Peas (in $\frac{1}{100}$ of an inch)	Mean Diameter of Offspring Peas
21	17.5
20	17.3
19	16.0
18	16.3
17	15.6
16	16.0
15	15.3

Source: Galton (1886).

Figure 1.2 display part of his data. For each of seven diameters, he found sweet peas having approximately that diameter and planted them. These were the "parent" sweet peas. After the plants grew and produced peas, these "offspring" sweet peas were harvested and their diameters were measured.

Galton noticed two things about these data. First, the averages of the offspring diameters had an approximately linear relationship with the parent diameters. Just by eye we can see in Figure 1.2 that a straight line could be drawn that fits the data fairly well.

Second, he noticed that the average diameters of the offspring peas appear to "regress" toward a common average. (In 1877 he used the word "revert" but in an 1885 paper he changed to the word "regress".) The overall

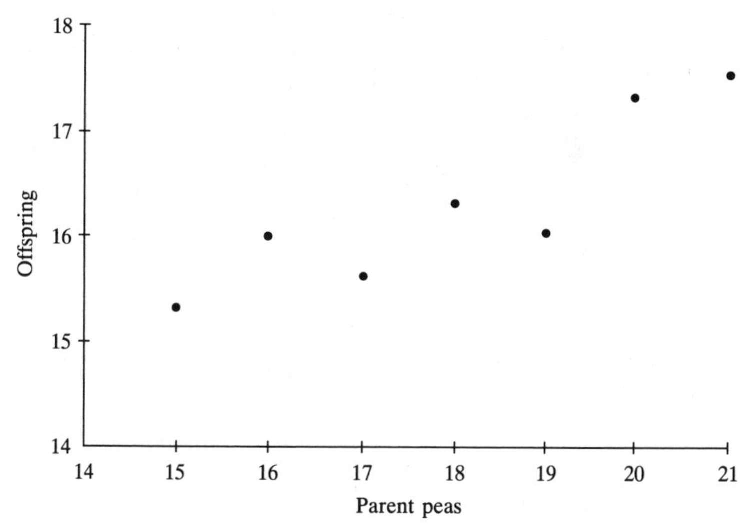

Figure 1.2 Plot of the sweet pea data.

average diameter of the offspring peas is about 16.3. For each of the seven parental diameters, the average diameter of the offspring differs from the parental diameter in the direction of the overall average. For example, the offspring of parents with diameter 21 have an average diameter of 17.5, which is in the direction of 16.3. And the offspring of parents with diameter 15 have an average diameter of 15.3, which is in the direction of 16.3. Galton later referred to this phenomenon as "regression toward mediocrity".

One might think that regression would imply that, after many generations, all sweet peas would end up having the same diameter. But the regression only pertains to the *average* diameters of the offspring peas. The individual peas have diameters that vary around the average. The variability of the individual diameters compensates for the regression of the average diameters, so that the distribution of diameters in the population of offspring peas is actually the same as the distribution of diameters in the population of parent peas.

NOTES

1.3a. In regression terminology we say that the variable Y "depends" on the variables $X_1, \ldots, X_p$ or that the value of Y is a "response" to the values of $X_1, \ldots, X_p$. Although the words suggest so, this should not be taken to mean that the relationship between the X's and Y is necessarily one of cause and effect. A regression model can be used to describe associative relationships as well as causative relationships.

1.3b. There are a number of reasons why we cannot expect Y to be an exact mathematical function of X_1, X_2, and X_3. The measurements of the three explanatory variables are incomplete in the sense that, for example, the air temperature measurement was taken at one particular time but the temperature probably varied somewhat throughout the production process. The measurements are subject to error due to imprecision in the measuring instruments and human errors. There are certainly other factors, besides sunlight, soil moisture, and air temperature, that influence the production of vitamin B_2. The production of vitamin B_2 is so complicated that it seems impossible to express it exactly in any mathematical formula.

1.7a. For more about Francis Galton's contributions to statistics see Stigler (1986, Chapter 8).

1.7b. To see how variability compensates for the regression of the averages, look at Figure 1.3. The ellipse represents a cloud of data points in a large sweet pea experiment. The x-coordinate of a data point is the diameter of a parent pea and the y-coordinate is the diameter of one of its offspring peas. Let us focus on peas having diameter 21 or more; for convenience we call these "large" peas. The large parent peas are those associated with the

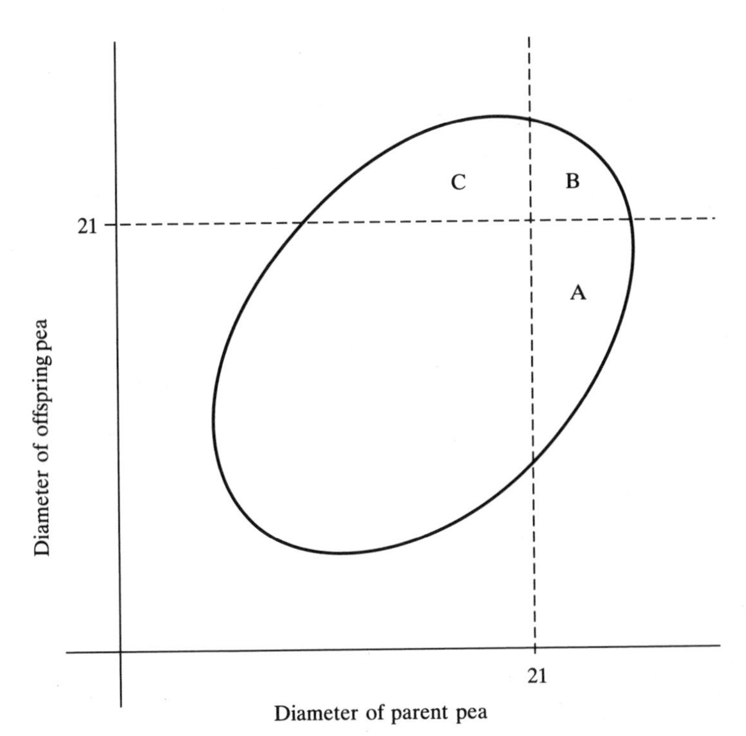

Figure 1.3 Ellipse representing a cloud of data points in a large sweet pea experiment.

points in areas A and B. Among the offspring of these large parent peas, only a fraction, due to the regression effect, are large; these are the offspring associated with the points in area B. But due to variability, some of the offspring of not-so-large parent peas are large; these are the offspring associated with the points in area C. So the proportion of large peas in the offspring population (associated with areas B and C) is the same as the proportion of large peas in the parent population (associated with areas A and B).

REFERENCES

Draper, N., and H. Smith (1981). *Applied Regression Analysis*. Wiley, New York.

Galton, F. (1877). Typical laws of heredity. *Nature*, vol. 15, pp. 492–495, 512–514, 532–533.

Galton, F. (1886). Regression towards mediocrity in hereditary stature. *Journal of the Anthropological Institute*, vol. 15, pp. 246–263.

Stigler, S. M. (1986). *The History of Statistics: The Measurement of Uncertainty Before 1900*. Harvard University Press, Cambridge, MA.

World Resources Institute (1986). *World Resources 1986*. Basic Books, New York.

CHAPTER 2

Constructing and Checking the Model

2.1 INTRODUCTION

The main subjects of this book are estimation and testing of regression coefficients, two basic components of regression analysis. In this chapter, before focusing on the various methods of estimation and testing presented in the remaining chapters, we first consider the overall plan of statistical analysis in which these methods are intended to be used.

Data analysis can be viewed as model construction. A model is proposed and then checked to see how well it fits the data. If the fit is inadequate, the model is modified. The process of checking and modifying the model is iterated until a suitable model is found.

Estimation and testing of regression coefficients are important tools in model construction. Note that we can choose to use any "brand" of tool, that is, any of the specific methods presented in Chapters 3 through 10.

Proposing a Model. It is sometimes the case that a model for the data has been suggested beforehand, either based on previous experience with similar data or based on theory about the process that generated the data. In the absence of such experience or theory, if we are given a set of regression data such as displayed in Table 1.2, it would be natural, as a first proposal, to try the model

$$Y = \beta_0 + \beta_1 X_1 + \beta_2 X_2 + \cdots + \beta_p X_p + e \tag{2.1}$$

Another strategy would be to look initially at a simple model such as $Y = \beta_0 + \beta_1 X_1 + e$ and then sequentially add other explanatory variables to the model.

The Assumptions. Whether the proposed model is valid or not depends on whether the data satisfy, at least approximately, the necessary assumptions. Recall the assumptions about the data that are implied by the linear regression model (2.1). In terms of the observed data the model is

$$y_i = \beta_0 + \beta_1 x_{i1} + \beta_2 x_{i2} + \cdots + \beta_p x_{ip} + e_i \tag{2.2}$$

for $i = 1, 2, \ldots, n$. The model assumes:

(i) Regarding y_i as a random variable, its expectation is a *function of the corresponding explanatory variables*, $x_{i1}, x_{i2}, \ldots, x_{ip}$, and the function is the *same* for all i.

(ii) The function in (i) is *linear*; that is, the expectation of y_i is $\beta_0 + \beta_1 x_{i1} + \beta_2 x_{i2} + \cdots + \beta_p x_{ip}$ for some coefficients β_j.

(iii) The "errors" $e_i = y_i - (\beta_0 + \beta_1 x_{i1} + \beta_2 x_{i2} + \cdots + \beta_p x_{ip})$ are like *independent random* draws from a *single* population.

This chapter describes some elementary procedures for checking the assumptions and for adjusting the model to achieve closer agreement and illustrates these procedures using both the least-squares method and an alternative regression method.

2.2 CHECKING THE MODEL

Plotting the Data. Suppose model (2.2) is proposed. A good first step in checking the model is to plot the response variable against each of the proposed explanatory variables; that is, for each j $(1 \le j \le p)$ plot the n points $(x_{1j}, y_1), (x_{2j}, y_2), \ldots, (x_{nj}, y_n)$. Such a plot may reveal violations of the model assumptions. If the cloud of points in the plot has a curved shape, this may indicate *nonlinearity* of the regression function, a violation of assumption (ii). If the cloud has a fan-like shape, this indicates *unequal variances* of the errors, a violation of assumption (iii). If one or more of the points stand apart from the main body of the cloud, these observations are possible *outliers*, which may not belong in the same model with the rest of the data.

Plotting the Residuals. If the p plots of Y against the explanatory variables do not reveal any obvious inadequacies in the model, the next step is to calculate estimates $\hat{\beta}_j$ for the regression coefficients. Any of the estimation methods described in Chapters 3 through 10 could be used. Then estimate the errors e_i by calculating the *residuals* $\hat{e}_i = y_i - (\hat{\beta}_0 + \hat{\beta}_1 x_{i1} + \hat{\beta}_2 x_{i2} + \cdots + \hat{\beta}_p x_{ip})$. Violations of the assumptions usually show up more clearly in plots of the residuals than in the data plots described in the preceding paragraph.

A *residual plot* is a plot of the n points $(z_1, \hat{e}_1), (z_2, \hat{e}_2), \ldots, (z_n, \hat{e}_n)$, where z_i is taken to be one of the explanatory variables x_{ij} or the fitted y-value $\hat{y}_i$ or any quantity that might possibly be related to the sign or size of the errors e_i. Sometimes the subscript i itself is related to e_i when the order of the subscripts is the same as the time sequence in which the data were collected. If the model is suitable, the residuals should behave approximately like the errors, which according to assumption (iii) should behave like independent random draws from a single population. The cloud of points in the residual plot should then be an amorphous cloud showing no pattern. Inspect the plot in the same way that the data plots were inspected, looking for nonlinearity, unequal variances, and outliers.

To make outliers easier to spot, it helps to divide the residuals by an estimate of their standard deviation and to plot these *standardized residuals*. A standardized residual is suspicious if it is larger than about 2.0 or 2.5 in absolute value.

Testing the Explanatory Variables. If the model seems adequate, we could then try to simplify it by eliminating explanatory variables that are only very weakly related to the response variable. If a certain variable or set of variables are suspected of being dispensible, we could test the hypothesis that their coefficients are zero.

2.3 MODIFYING THE MODEL

When we apply the procedures of the preceding section to check the model, we may find that the model fits the data well. If not, we can try to modify the model to achieve a better fit.

Nonlinearity. Suppose the data plots or the residual plots reveal that the relationship between the response variable and the explanatory variables is nonlinear. We can try changing the variables in some way so that the relationship becomes approximately linear. The plots may suggest a particular transformation of a variable. For example, if the plot of Y versus X_1 looks like the right half of a parabola, then we could try replacing Y by $\sqrt{Y}$ or replacing X_1 by X_1^2. Sometimes the nature of a variable suggests a transformation. If Y is the efficiency of a car, measured in miles per gallon, the transformed variable $1/Y$, in gallons per mile, makes equally good sense as a measure of inefficiency. Or if Y is the length of a fish, the transformed variable Y^3 could be interpreted as a proxy for the weight of the fish.

Another way to achieve approximate linearity is to add explanatory variables to the model. These could be entirely new variables or variables such as X_1^2 or X_1X_2, which are constructed from other variables.

Unequal Variances. If the plots indicate unequal variances of the errors, a transformation of Y may help. If Y is always positive and its range spans several orders of magnitude, such as from 0.01 to 10, a frequently used transformation is the logarithm function. If Y is a percentage, a commonly used transformation for stabilizing its variance is the arcsine square-root function.

Outliers. If a data point stands apart from the main body of points in a plot, it should be investigated to see what may have caused this. Perhaps a number has been incorrectly recorded. Or the experimenter who collected the data may remember some special circumstance for that data point. If special factors have influenced an outlier, one could remove it from the data set and use a linear regression model only for the remaining data. (The information that certain factors may produce outliers should be retained and could even be the most interesting result of the analysis.) However, an outlier may legitimately belong in the same model with the other data; it may be that the process that generated the data is one that occasionally produces extreme values. When an outlier is kept in the data set, its effect on the analysis can be checked by running two analyses, one with and one without the outlier. The outlier's effect can be constrained by using least-absolute-deviations, M-, or nonparametric regression.

2.4 EXAMPLES

Since there is always a lot of subjectivity involved in planning and interpreting a statistical analysis, each of the following examples should be viewed as only one of many possible analyses.

Example 1. Consider the turnip green data described in Section 1.2 and displayed in Table 1.1. Plots of the response variable, Y = vitamin B_2, versus each of the three explanatory variables, X_1 = sunlight, X_2 = soil moisture, and X_3 = air temperature, are shown in Figure 2.1.

The plot of Y versus X_2 looks nonlinear. A corrective measure for this is to add X_2^2 to the model as a fourth explanatory variable. That is, let us modify the model to be

$$Y = \beta_0 + \beta_1X_1 + \beta_2X_2 + \beta_3X_3 + \beta_4X_4 + e \tag{2.3}$$

where $X_4 = X_2^2$.

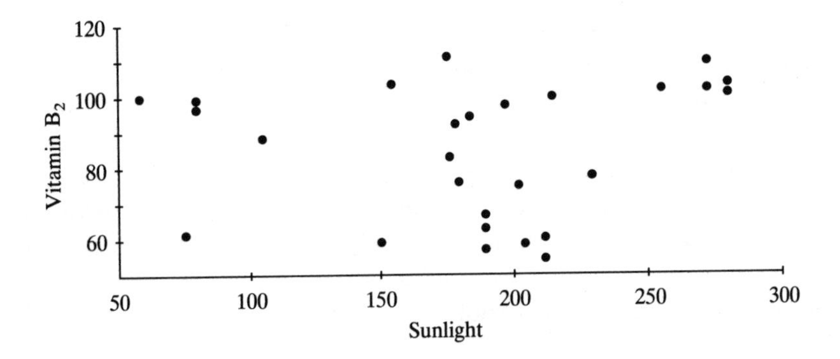

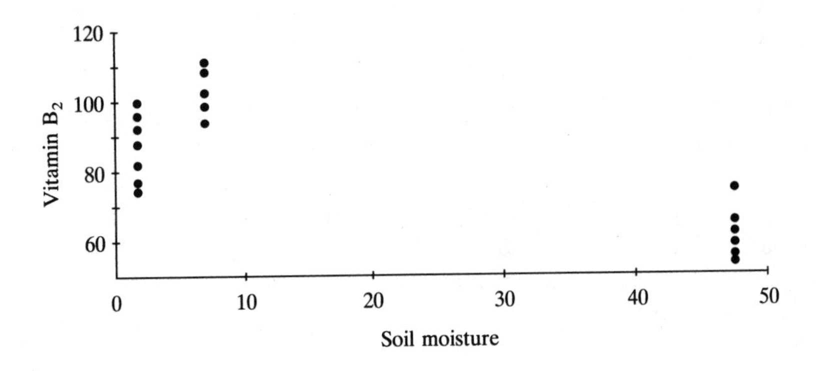

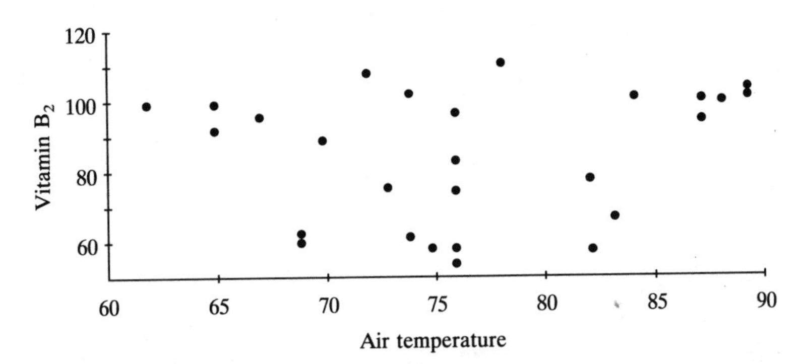

Figure 2.1 Data plots for the turnip data.

Now we want to estimate the coefficients in this model. Suppose we decide to use the method of least squares. Following the procedure described in Chapter 3, we obtain $\hat{Y} = 119.6 - 0.03367X_1 + 5.425X_2 - 0.5026X_3 - 0.1209X_2^2$ as an estimate of the regression equation. The least-squares estimate of the standard deviation of the errors is $\hat{\sigma} = 6.104$.

Next we calculate the residuals $\hat{e}_i = y_i - \hat{y}_i$ and plot their standardized values $\hat{e}_i/\hat{\sigma}$ versus the fitted values $\hat{y}_i$ and the three explanatory variables x_{i1}, x_{i2}, and x_{i3}. These four plots are shown in Figure 2.2.

No curvature is apparent in these plots. The plot versus X_2, soil moisture, emphasizes that this variable takes on only three different values—2.0, 7.0, and 47.4. We would have to be cautious about the validity of the estimated regression equation for data with soil-moisture measurements of 20 or 30. This plot also suggests that the variance of the errors for $X_2 = 7.0$ may be smaller than for the other two values of X_2. But the observed difference in variance could simply be due to chance since the difference is caused by only two or three points with large residuals.

A standardized least-squares residual can be regarded as a possible outlier, to be investigated, if its absolute value is greater than 2.0. (Some statisticians would not worry unless the absolute value was greater than 2.5.) The plots show one possible outlier, plant number 19, which has a standardized residual of 2.286. In the absence of a specific reason for omitting it, we retain it in the model. Moreover, its residual is not too extreme, and among more than 20 observations we expect to see at least one observation whose standardized residual is larger than 2 in absolute value.

Thus we see that linear regression model (2.3) seems to adequately fit these data. Now we check whether the model can be simplified by dropping one or more explanatory variables. Using the procedure presented in Chapter 3, we can test the four hypotheses that $\beta_1 = 0$, $\beta_2 = 0$, $\beta_3 = 0$, and $\beta_4 = 0$. The test of $\beta_1 = 0$ indicates that perhaps $\beta_1 = 0$, and hence that explanatory variable X_1, the amount of sunlight, does not contain much information about the response variable beyond the information already contained in the other explanatory variables.

Let us simplify the model to

$$Y = \beta_0 + \beta_2 X_2 + \beta_3 X_3 + \beta_4 X_2^2 + e \tag{2.4}$$

The least-squares estimate of the regression equation is $\hat{Y} = 120.6 + 4.904X_2 - 0.5716X_3 - 0.1108X_2^2$ and the estimate of the standard deviation of the errors is $\hat{\sigma} = 6.223$. Testing the three hypotheses $\beta_2 = 0$, $\beta_3 = 0$, and $\beta_4 = 0$ in this model, we find that all three explanatory variables are necessary. The residual plots for this model look similar to the corresponding plots in Figure 2.2 for model (2.3), with the slight difference that, in the simplified

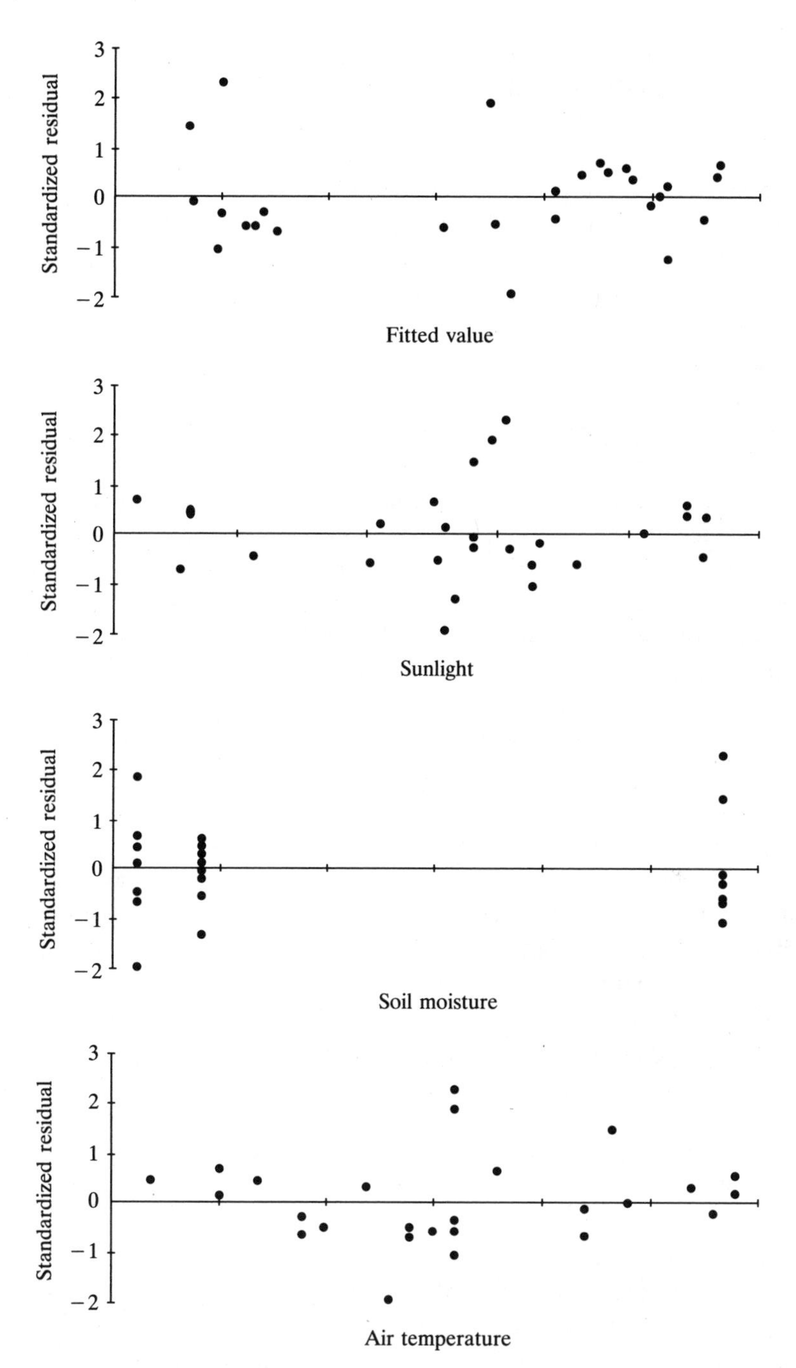

Figure 2.2 Standardized least-squares residual plots for the turnip data with the added explanatory variable $X_4 = X_2^2$.

model, plant number 15, in addition to plant number 19, now qualifies as an outlier with a standardized residual of -2.084. Model (2.4) seems to be a satisfactory model for the turnip data.

Example 2. Consider the fire data described in Section 4.5 and displayed in Table 4.3. The response variable, $Y =$ the incidence of fires, is plotted versus each of three explanatory variables, $X_1 =$ the proportion of older housing units, $X_2 =$ the incidence of theft, and $X_3 =$ the median family income, in Figure 2.3.

In the plot of Y versus X_2 we see one point that is far to the right of the other points. The incidence of theft in area number 24 is 147, whereas the incidence of theft in each of the other areas is no more than 75. A data point that has an extreme value for one of the explanatory variables is called a *leverage point*. (The term *outlier* is usually reserved for data points that have extreme values for the response variable.) Leverage points have a disproportionate influence on the estimates of the regression coefficients, which is especially undesirable because it often happens that leverage points do not follow the same model as the rest of the data. Therefore we delete area 24 from the model, and confine our conclusions to areas with incidence of theft no more than 75. The plot of Y versus X_3 shows that area number 7 is also a leverage point. Its median family income is 21,480 whereas the next largest value is only 16,250 and the remaining values are all less than 14,000. We will also delete area 7 from the model.

In the plot of Y versus X_1 the cloud of points has a fan-like shape, indicating unequal variances. The plot of Y versus X_3 also as a fan-like shape, with the fan opening in the opposite direction. Let us try to correct this problem by transforming Y by the logarithm function. (Logarithms to the base e are used here but the base is not important.) Plots of $\log Y$ versus each of the explanatory variables are shown in Figure 2.4.

Since these plots do not show any obvious violations of the assumptions, let us try the linear regression model

$$\log Y = \beta_0 + \beta_1 X_1 + \beta_2 X_2 + \beta_3 X_3 + e \tag{2.5}$$

Let us use the method of least absolute deviations (LAD) to estimate the regression coefficients. Following the procedure of Chapter 4, we obtain $\hat{Y} = 4.362 - 0.09098X_1 + 0.01299X_2 - 0.2425X_3$. The LAD estimate of the standard deviation of the errors is $\hat{\sigma} = 0.5100$.

The standardized residuals $\hat{e}_i/\hat{\sigma}$ are plotted versus the fitted values $\hat{y}_i$ and each of the three explanatory variables in Figure 2.5.

These plots look acceptable with respect to the assumptions of linearity and equal variance. The standardized residuals of the areas numbered 13 and

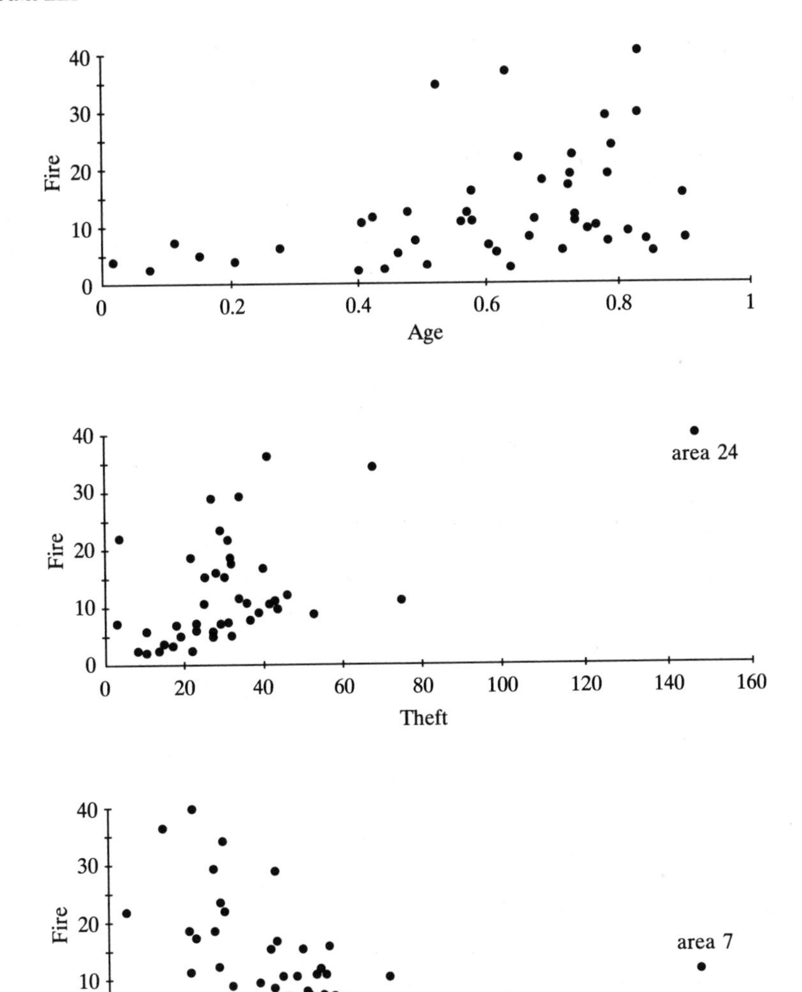

Figure 2.3 Data plots for the fire data.

34 are greater than 2 but less than 2.5. LAD regression generally produces larger standardized residuals than least-squares regression, so a value between 2 and 2.5 is not as suspect. Area 13 has the second largest median family income, and so one could justifiably label it as a leverage point and delete it from the model, but we have kept it.

At this point, model (2.5) seems like a reasonably good model for the fire data with areas 7 and 24 omitted. Let us see if it can be simplified. Using the procedure presented in Chapter 4, we can test the three hypotheses that

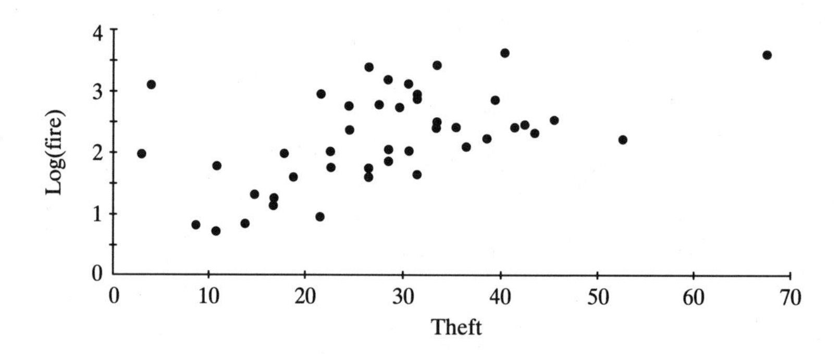

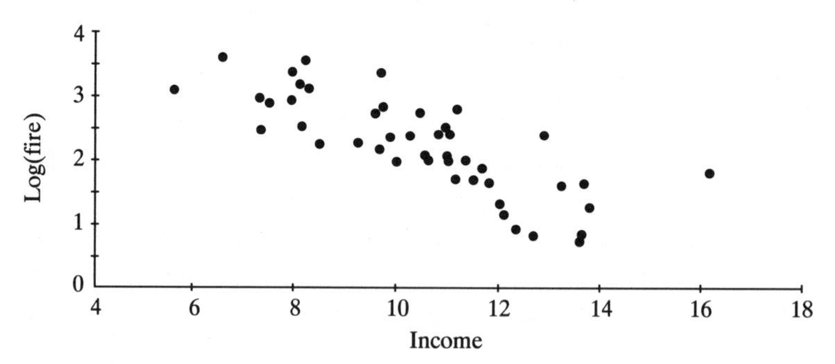

Figure 2.4 Data plots for the fire data after deleting areas 7 and 24 and transforming Y to $\log Y$.

$\beta_1 = 0$, $\beta_2 = 0$, and $\beta_3 = 0$. The test of the first hypothesis leads us to drop explanatory variable X_1, the proportion of older housing units, from the model.

Consider the model

$$\log Y = \beta_0 + \beta_2 X_2 + \beta_3 X_3 + e$$

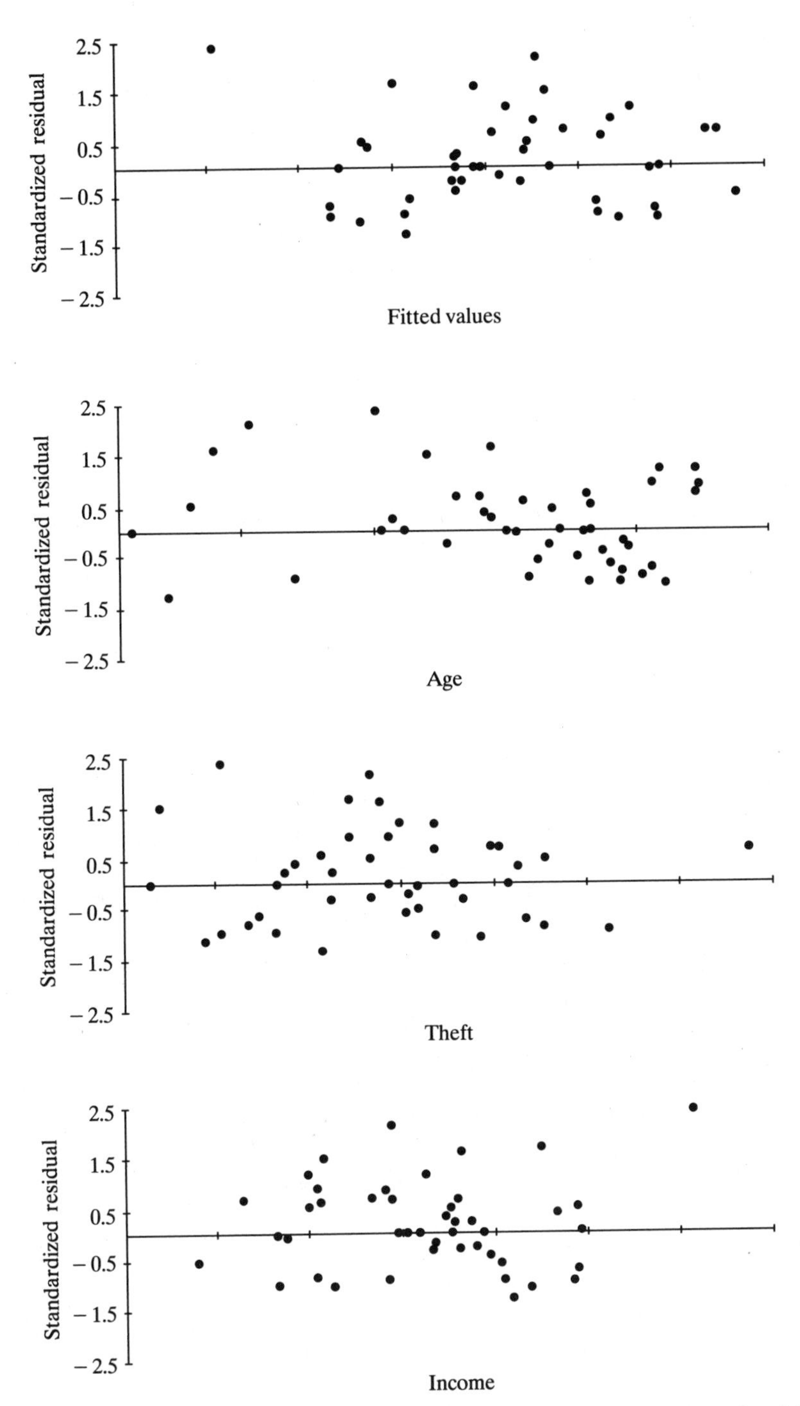

Figure 2.5 Standardized least-absolute-deviations residual plots for the fire data after deleting areas 7 and 24 and transforming Y to $\log Y$.

The LAD estimate of the regression equation is $\hat{Y} = 4.252 + 0.01156X_2 - 0.2330X_3$ and $\hat{\sigma} = 0.5232$. Testing the two hypotheses $\beta_2 = 0$ and $\beta_3 = 0$ in this model, we find that we could try removing X_2 from the model.

Now consider the model

$$\log Y = \beta_0 + \beta_3 X_3 + e \tag{2.6}$$

The LAD estimate of the regression equation is $\hat{Y} = 4.988 - 0.2718X_3$ and $\hat{\sigma} = 0.5325$. The standardized residual plots look similar to those versus $\hat{Y}$ and X_3 in Figure 2.5. There is no evidence of nonlinearity or nonconstant variance or outliers. Model (2.6) appears to be a satisfactory model for the fire data.

NOTES

2.1a. The process of model construction is fundamental to the scientific method. See Box, Hunter, and Hunter (1978, Chapter 1).

2.1b. In this chapter we have described the process of model construction as a sequential process in which a sequence of models are considered, one at a time, but other approaches are possible. In the examples in Section 2.4 we have followed a particular sequential procedure called backward elimination. One begins with model (2.1), containing all available explanatory variables, and then sequentially eliminates those variables that are unnecessary. Hypothesis testing is used to determine which variables are unnecessary.

Other sequential strategies for model construction are forward selection, mentioned in the last sentence of the subsection on proposing a model, and the stepwise method, in which explanatory variables are added at some steps and deleted at other steps.

The all-possible-regressions procedure is a nonsequential approach. To use this procedure one must choose a measure of the goodness of a model, such as Mallow's C_p criterion (see Weisberg, 1985, p. 216, or Myers, 1990, p. 182). The criterion is calculated for all the linear regression models obtained by using all possible subsets of the p available explanatory variables. Among these 2^p models, one could choose the model with the best value of the criterion. Or one might choose several models with good values of the criterion and investigate them in more detail. Of course, the feasibility of looking at all possible regressions depends on the size of p.

See the topic of *variable selection* in the books suggested for additional reading at the end of the Notes in Chapter 3.

2.2a. Methods for checking the assumptions of a model are often called *diagnostics*. Most of these methods involve analysis of residuals.

2.2b. See Note 2.4a for more about standardization of residuals.

2.3a. A transformation of the data can be used to improve their agreement with the linear regression model. A suitable transformation of Y can sometimes be estimated from the data. The Box–Cox family of power transformations can be used for this purpose. One supposes that Y^λ follows a linear regression model for some power λ, and then uses the data to estimate the value of λ. This family of transformations includes $1/Y$, $\sqrt{Y}$, Y, Y^2, Y^3, and $\log Y$. (For $\lambda = 0$ we let $Y^\lambda = \log Y$, which is sensible because $\log Y$ is the limit of $(Y^\lambda - 1)/\lambda$ as λ approaches 0.) See Draper and Smith (1981, Section 5.3), Weisberg (1985, Sections 6.4 and 6.5), or Myers (1990, Section 7.3).

2.3b. A linear transformation of the data has no effect on the data's agreement with the linear regression model. Suppose we transform Y to $Z = cY + d$. The model $Z = \gamma_0 + \gamma_1 X_1 + \cdots + \gamma_p X_p + e$ is equivalent to the model $Y = \beta_0 + \beta_1 X_1 + \cdots + \beta_p X_p + e$, where $\beta_0 = (\gamma_0 - d)/c$, $\beta_1 = \gamma_1/c, \ldots, \beta_p = \gamma_p/c$. Or suppose we transform X_1 to $W = cX_1 + d$. The model $Y = \gamma_0 + \gamma_1 W + \gamma_2 X_2 + \cdots + \gamma_p X_p + e$ is equivalent to the model $Y = \beta_0 + \beta_1 X_1 + \beta_2 X_2 + \cdots + \beta_p X_p + e$, where $\beta_0 = \gamma_0 + d\gamma_1$, $\beta_1 = c\gamma_1$, $\beta_2 = \gamma_2, \ldots, \beta_p = \gamma_p$.

Therefore a linear change in units, such as from miles to kilometers or from degrees Fahrenheit to degrees Celsius, has no essential effect on a linear regression model. Results from a regression analysis using the changed model can easily be translated back in terms of the original model. A nonlinear change in units, however, such as from miles per gallon to gallons per mile or from centimeters to cubic centimeters, produces an essentially different linear regression model. Results from a regression analysis using the changed model do not necessarily imply anything about the original model.

2.3c. A discussion of the frequency of occurrence of outliers (or gross errors) in real data is given in Section 1.2c of Hampel et al. (1986).

2.4a. A residual is *standardized* by dividing by an estimate of its standard deviation. In Examples 1 and 2 we have standardized $\hat{e}_i$ by dividing it by $\hat{\sigma}$. This is sensible because the residuals are estimates of the random errors, and so $\text{SD}(\hat{e}_i) \approx \text{SD}(e_i) = \sigma$. But a more accurate estimate of $\text{SD}(\hat{e}_i)$ can be obtained.

For least-squares residuals, $\text{SD}(\hat{e}_i) = \sigma\sqrt{1 - h_{ii}}$, where h_{ii} is the ith diagonal entry in the matrix $\boldsymbol{X}(\boldsymbol{X}'\boldsymbol{X})^{-1}\boldsymbol{X}'$ and $\boldsymbol{X}$ is the $n \times (p + 1)$ matrix of explanatory variables x_{ij} augmented by a column of 1's. Hence a more refined definition of a standardized least-squares residual would be $\hat{e}_i/(\hat{\sigma}\sqrt{1 - h_{ii}})$, rather than $\hat{e}_i/\hat{\sigma}$. This is sometimes called a *studentized* residual. In Example 1, there are three turnip plants having studentized residuals with absolute value greater than 2. The plants numbered 10, 15, and 19 have studentized residuals 2.077, -2.093, and 2.435, respectively.

It can be shown that h_{ii} is always between 0 and 1; hence $\sqrt{1 - h_{ii}}$ is less than 1, and the studentized least-squares residual is always larger than the standardized least-squares residual. In simple least-squares regression, $h_{ii} = (1/n) + (x_i - \bar{x})^2/\Sigma(x_i - \bar{x})^2$, and so the difference between the studentized residual and the standardized residual is substantial only when n is small and x_i is far from $\bar{x}$.

For studentization of residuals in M-regression, see Belsley, Kuh, and Welsch (1980, p. 234). For studentization of residuals in nonparametric regression, see McKean, Sheather, and Hettmansperger (1990, Section 4).

2.4b. The problem of leverage points is covered in Belsley, Kuh, and Welsch (1980, Chapter 2), Cook and Weisberg (1982), and Rousseeuw and Leroy (1987, Chapter 6). Briefer treatments of the problem can be found in introductory regression books such as Draper and Smith (1981, Section 3.12), Weisberg (1985, Section 5.3), and Myers (1990, Chapter 6).

Discussion of leverage points may often be found under the topic of "influential observations". A data point is an *influential observation* if it has a large influence on an estimate or test. In particular, consider the least-squares estimate $\hat{y}_i = \hat{\beta}_0 + \hat{\beta}_1 x_{i1} + \cdots + \hat{\beta}_p x_{ip}$ of the regression equation at the x-values of the ith data point. It can be expressed as $\hat{y}_i = (1 - h_{ii})\hat{y}_i^* + h_{ii} y_i$ where $\hat{y}_i^*$ is the estimate obtained without using the ith data point and h_{ii} is as in Note 2.4a. The quantity h_{ii} is called the *potential* of the ith data point. The data point is influential if either h_{ii} is large (which happens for leverage points) and y_i is moderately different from $\hat{y}_i^*$ or if y_i is much different from $\hat{y}_i^*$ (which happens for outliers) and h_{ii} is not too small. A typical value of the potential h_{ii} is about $(p + 1)/n$; a potential is usually regarded as "large" if it is larger than $2(p + 1)/n$.

In the fire data of Example 2, the potential values of areas 7 and 24 are 0.5200 and 0.6173, which are much larger than $2(p + 1)/n = 2(3 + 1)/47 = 0.1702$.

2.4c. When transforming Y by the logarithm function, it is not important whether you use logarithms to the base e or base 10 or some other base. We have used $\log_e Y$ but it would be essentially equivalent to use $\log_{10} Y$ because $\log_{10} Y = c \log_e Y$, where c is the constant $\log_{10} e$. As seen in Note 2.3b, multiplication of a variable by a constant has no essential effect in linear regression.

2.4d. Figure 2.5 includes a plot of LAD residuals versus fitted y-values. Such plots cannot necessarily be interpreted in the same way as plots of least-squares residuals versus fitted y-values, because in LAD regression the residuals can be negatively correlated with the fitted y-values; see Sheather and McKean (1992). Therefore a pattern in the plot could be due simply to this correlation rather than any inadequacy in the model. The correlation should be no worse than about $-\sqrt{(p+1)}/n$. For $p = 3$ and $n = 45$, as in the fire data of Example 2, the correlation should no worse than about -0.05.

Additional Reading. More on checking and constructing models can be found in the books that are suggested as additional reading for Chapter 3. For a more comprehensive coverage of regression diagnostics see the books by Belsley, Kuh, and Welsch (1980), Cook and Weisberg (1982), or Atkinson (1985). These books focus on the method of least squares. Diagnostic procedures have been developed for nonparametric rank-based regression by McKean, Sheather, and Hettmansperger (1990). Robust diagnostics are given in Rousseeuw and Leroy (1987, Section 6.6). The problem of outliers is treated in books by Barnett and Lewis (1984) and Hawkins (1980). Also see Rousseeuw and Leroy (1987, Chapter 6) and Beckman and Cook (1983, Section 4). The use of transformations is treated in books by Atkinson (1985) and Carroll and Ruppert (1988).

REFERENCES

Atkinson, A. C. (1985). *Plots, Transformations, and Regression: An Introduction to Graphical Methods of Diagnostic Regression Analysis*. Oxford University Press, Oxford.

Barnett, V., and T. Lewis (1984). *Outliers in Statistical Data*, 2nd ed. Wiley, Chichester.

Beckman, R. J., and R. D. Cook, (1983). Outlier..........s. *Technometrics*, vol. 25, pp. 119–149.

Belsley, D. A., E. Kuh, and R. E. Welsch (1980). *Regression Diagnostics: Identifying Influential Data and Sources of Collinearity*. Wiley, New York.

Box, G. E. P., W. G. Hunter, and J. S. Hunter (1978). *Statistics for Experimenters: An Introduction to Design, Data Analysis, and Model Building*. Wiley, New York.

Carroll, R. J., and D. Ruppert (1988). *Transformation and Weighting in Regression*. Chapman and Hall, New York.

Cook, R. D., and S. Weisberg (1982). *Residuals and Influence in Regression*. Chapman and Hall, New York.

Draper, N. R., and H. Smith (1981). *Applied Regression Analysis*, 2nd ed. Wiley, New York.

Hampel, F. R., E. M. Ronchetti, P. J. Rousseeuw, and W. A. Stahel (1986). *Robust Statistics: The Approach Based on Influence Functions*. Wiley, New York.

Hawkins, D. M. (1980). *Identification of Outliers*. Chapman and Hall, London.

McKean, J. W., S. J. Sheather, and T. P. Hettmansperger (1990). Regression diagnostics for rank-based methods. *Journal of the American Statistical Association*, vol. 85, pp. 1018–1028.

Myers, R. H. (1990). *Classical and Modern Regression with Applications*. PWS-Kent, Boston.

Rousseeuw, P. J., and A. M. Leroy (1987). *Robust Regression and Outlier Detection*. Wiley, New York.

Sheather, S. J., and J. W. McKean (1992). The interpretation of residuals based on L_1 estimation. In: Y. Dodge (ed.), L_1-*Statistical Analysis and Related Methods*. North-Holland, New York.

Weisberg, S. (1985). *Applied Linear Regression*, 2nd ed. Wiley, New York.

CHAPTER 3

Least-Squares Regression

3.1 INTRODUCTION

The most commonly used regression method is the method of least squares. It was discovered independently by Carl Friedrich Gauss in Germany around 1795 and by Adrien Marie Legendre in France around 1805. Early applications of the method were to astronomic and geodetic data. Its first published appearance was in 1805 in an appendix to a book by Legendre on determining the orbits of comets.

3.2 AN EXAMPLE OF SIMPLE REGRESSION

An experiment was conducted to find the relationship between two procedures for assessing the acid content of a chemical. The two procedures measure somewhat different but related quantities. The organic acid content of a sample of chemical can be determined by a method of extraction and weighing, which is expensive, but a relatively cheap titration method is available for determining the acid number. It was hoped that, by using regression, the cheap method could be used instead of the expensive method to measure organic acid content.

Both procedures were used on 20 samples of chemical. The data are displayed in Table 3.1. Using these data we would like to obtain an equation that expresses the organic acid content measurement as an approximate function of the acid number measurement. For notation, let

y_i = expensive organic acid content measurement of the ith chemical sample

x_i = cheap acid number measurement of the ith chemical sample

The data points $(x_1, y_1), \ldots, (x_{20}, y_{20})$ are plotted in Figure 3.1. The relationship between X and Y appears to be approximately linear.

Table 3.1 Acid Content Data

Identification Number of Chemical Sample	Expensive Measurement (Y)	Cheap Measurement (X)
1	76	123
2	70	109
3	55	62
4	71	104
5	55	57
6	48	37
7	50	44
8	66	100
9	41	16
10	43	28
11	82	138
12	68	105
13	88	159
14	58	75
15	64	88
16	88	164
17	89	169
18	88	167
19	84	149
20	88	167

Source: Daniel and Wood (1980, p. 46).

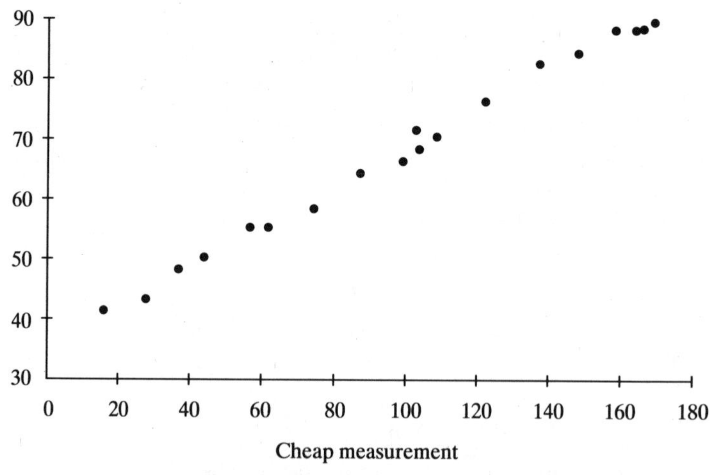

Figure 3.1 Plot of the acid content data.

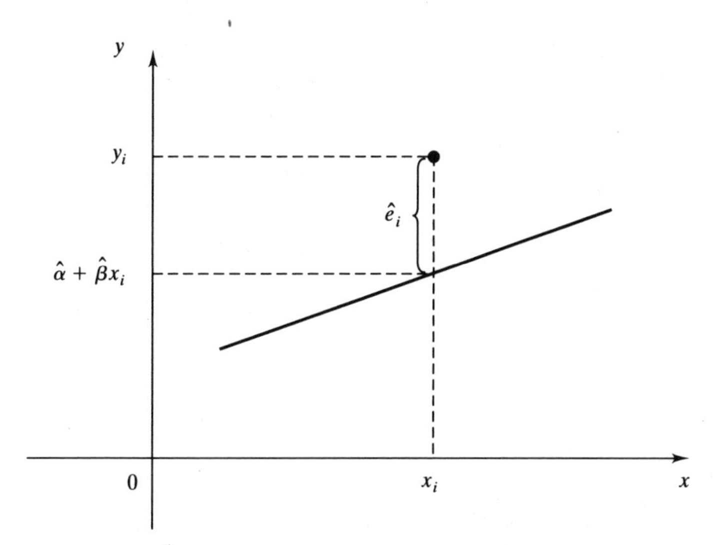

Figure 3.2 The residual $\hat{e}_i$ is the vertical distance from the estimated regression line to the data point (x_i, y_i).

Let us try the linear regression model given in equation (1.1). For this example, the value of p is 1. That is, we have only one explanatory variable, in which case the model is called a *simple* linear regression model. Strictly following equation (1.1), one would write $Y = \beta_0 + \beta_1 X_1 + e$, but to avoid unnecessary subscripts, let us write the model as

$$Y = \alpha + \beta X + e \tag{3.1}$$

Next we should estimate α and β. Then, if we use the cheap procedure to measure the acid number X of a chemical sample, we can estimate its organic acid content to be $\hat{Y} = \hat{\alpha} + \hat{\beta}X$.

3.3 ESTIMATING THE REGRESSION LINE

There are various methods one can use to estimate α and β in equation (3.1). This chapter presents the least-squares method.

The basic idea in estimating the regression line is to find the straight line that "best" fits the data points in the plot. To judge how well the estimated regression line $\hat{Y} = \hat{\alpha} + \hat{\beta}X$ fits the data, we can look at the size of the residuals $\hat{e}_i = y_i - (\hat{\alpha} + \hat{\beta}x_i)$. Note in Figure 3.2 that the residual $\hat{e}_i$ is the vertical distance from the estimated regression line to the data point (x_i, y_i). We want to choose $\hat{\alpha}$ and $\hat{\beta}$ so that the residuals are "small".

In the least-squares method of regression, the overall size of the residuals is measured by $\Sigma\hat{e}_i^2$. The *least-squares estimates* of α and β are defined to be the values $\hat{\alpha}$ and $\hat{\beta}$ which give the smallest value of $\Sigma\hat{e}_i^2$, that is, which give the least sum of squares of the residuals. Formulas for the least-squares estimates are:

$$\hat{\beta} = \frac{\Sigma(x_i - \bar{x})(y_i - \bar{y})}{\Sigma(x_i - \bar{x})^2}$$
$$\hat{\alpha} = \bar{y} - \hat{\beta}\bar{x} \tag{3.2}$$

Here $\bar{x}$ and $\bar{y}$ denote the averages of the x_i's and the y_i's, respectively.

Reasonableness of the Formulas. Before giving a formal justification of these formulas, let us reexpress them in a way that makes them appear "reasonable". The formula for $\hat{\alpha}$ can be rewritten as

$$\bar{y} = \hat{\alpha} + \hat{\beta}\bar{x}$$

This says that the estimated regression line passes through the point $(\bar{x}, \bar{y})$, which can be regarded as the center of the cloud of data points. This is a good property for a line that is intended to "fit" the data.

The formula for $\hat{\beta}$ can be rewritten as

$$\hat{\beta} = \sum w_i \left(\frac{y_i - \bar{y}}{x_i - \bar{x}} \right), \qquad w_i = \frac{(x_i - \bar{x})^2}{\Sigma(x_i - \bar{x})^2}$$

Note that $(y_i - \bar{y})/(x_i - \bar{x})$ is the slope of the line between the central point $(\bar{x}, \bar{y})$ and the data point (x_i, y_i). Thus the slope $\hat{\beta}$ of the estimated regression line is a kind of average of these slopes. More precisely, it is a *weighted average*. Note that the weights w_i are nonnegative and their sum is 1. (An ordinary average is the special case of a weighted average in which all the weights are equal to $1/n$.) Each data point is weighted proportionally to the square of its x-distance from the center. This implies that data points far from the center of the data have a large influence on the estimate of the slope of the regression line.

Formal Justification of the Formulas. Suppose a and b are candidates for the regression estimates $\hat{\alpha}$ and $\hat{\beta}$. Using the line $\hat{Y} = a + bX$, the sum of squares of the residuals is $\Sigma(y_i - a - bx_i)^2$. Considering this as a function of a and b, its minimum can be found by using calculus. Take the partial derivatives with respect to a and b, and solve for the values of a and b that

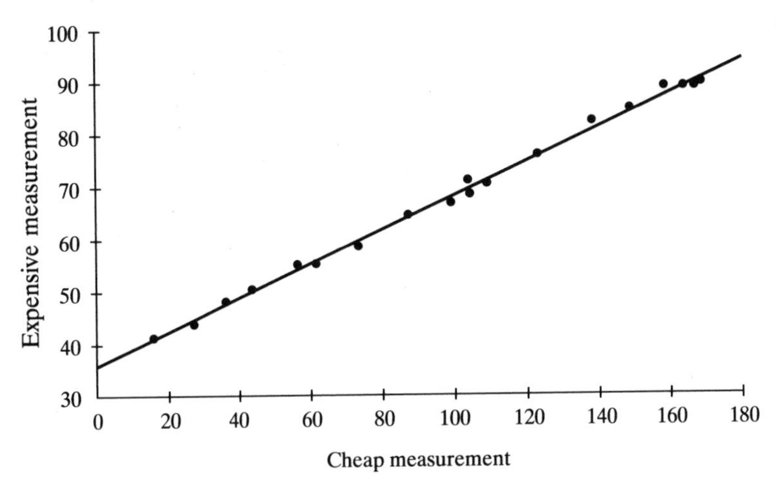

Figure 3.3 Plot and least-squares regression line for the acid content data.

make these partial derivatives equal to 0. The solutions are $a = \hat{\alpha}$ and $b = \hat{\beta}$, where $\hat{\alpha}$ and $\hat{\beta}$ are given by (3.2).

The Acid Content Data. For the data in Table 3.1 the formulas in (3.2) yield the estimates $\hat{\alpha} = 35.46$ and $\hat{\beta} = 0.3216$. So the fitted regression line is $\hat{Y} = 35.46 + 0.3216X$. This is the line in Figure 3.3.

3.4 TESTING $\beta = 0$

In Figure 3.3 it seems clear that the explanatory variable X has a close relationship with the response variable Y. More formally, the significance of the relationship can be tested as follows. We can compare the full model $Y = \alpha + \beta X + e$ using X and the reduced model $Y = \alpha + e$ without X. If the full model fits the data significantly better than the reduced model, this implies that the relationship between X and Y is significant. Another way of describing the comparison of the two models is to call it a test of the hypothesis that $\beta = 0$.

Based on the evidence of the data, we must decide whether $\beta = 0$ or $\beta \neq 0$. The hypothesis $\beta = 0$ (the *null hypothesis*) and the hypothesis $\beta \neq 0$ (the *alternative hypothesis*) are not treated equally. Because of our preference for simplicity, even before we look at the evidence we begin with a preference for $\beta = 0$, because the model $Y = \alpha + e$ is simpler than the model $Y = \alpha + \beta X + e$. So our attitude is that we will decide $\beta = 0$ unless there is strong evidence against it.

Description of the Test. First we describe how the least-squares test of $\beta = 0$ is performed. Justification of the procedure is given afterward.

To test $\beta = 0$, first use (3.2) to calculate $\hat{\beta}$, and then calculate

$$\hat{\sigma} = \sqrt{\frac{\Sigma \hat{e}_i^2}{n-2}} \tag{3.3}$$

Then calculate an estimate of the standard deviation of $\hat{\beta}$ by substituting $\hat{\sigma}$ for σ in the formula

$$\mathrm{SD}(\hat{\beta}) = \frac{\sigma}{\sqrt{\Sigma (x_i - \bar{x})^2}} \tag{3.4}$$

where σ denotes the standard deviation of the population of errors. The test statistic is

$$|t| = \frac{|\hat{\beta}|}{\mathrm{est.SD}(\hat{\beta})} \tag{3.5}$$

(We use absolute value notation for $|t|$ because later we talk about $t = \hat{\beta}/\mathrm{est.SD}(\hat{\beta})$.)

The p-value of the test is obtained from the t distribution with $n - 2$ degrees of freedom. The p-value is calculated as the probability that the absolute value of a random variable with this distribution is greater than or equal to the value of $|t|$ calculated from (3.5).

Justification of the Test. The most relevant information we have about β is the estimate $\hat{\beta}$. The value of $\hat{\beta}$ should indicate whether $\beta = 0$ or not. There is strong evidence that $\beta \neq 0$ when $\hat{\beta}$ is "far" from 0.

For the acid content data we calculated $\hat{\beta} = 0.3216$. Is this far from 0? It depends on how variable $\hat{\beta}$ is. When we speak of the variability of $\hat{\beta}$, we are thinking of $\hat{\beta}$ as a random variable. Imagine repeating the acid content experiment an infinite number of times. The infinite number of values of $\hat{\beta}$ obtained in these experiments would be centered around β but they would vary. The size of a typical deviation of $\hat{\beta}$ from β is measured by the standard deviation of $\hat{\beta}$. We can estimate $\mathrm{SD}(\hat{\beta})$ by substituting (3.3) into (3.4) to obtain $\mathrm{est.SD}(\hat{\beta})$. For the acid content data, $\mathrm{est.SD}(\hat{\beta}) = 0.0056$. So the distance between $\hat{\beta} = 0.3216$ and 0 is about 57 ($= 0.3216/0.0056$) times the size of $\mathrm{SD}(\hat{\beta})$. This makes it very unlikely that $\beta = 0$.

Thus we see that a reasonable test of $\beta = 0$ can be based on the test statistic $|t|$ in (3.5). A very large value of $|t|$ means that $\hat{\beta}$ is much farther from 0 than would be expected if $\beta = 0$. This constitutes strong evidence against $\beta = 0$.

Reasonableness of the Formula for SD($\hat{\beta}$). Consider formula (3.4). It implies that SD($\hat{\beta}$) is smaller when σ is smaller, that is, when the variability of the random errors is smaller. This makes sense since we should be able to estimate β more accurately in the presence of smaller errors. Also, SD($\hat{\beta}$) is smaller when $\Sigma(x_i - \bar{x})^2$ is larger, that is, when the x-values are more spread out. This also makes sense if we think about the following analogy. Suppose we want to use a pencil and ruler to draw a horizontal line 10 centimeters above the bottom of a sheet of paper. We could draw two points 10 centimeters above the bottom and then align the ruler on these two points and draw a line. Because of human error, the line would not be exactly horizontal. To get the line to be as horizontal as possible, we would choose the two points to be on opposite edges of the sheet of paper, that is, as spread out as possible.

Estimating σ. A natural estimate of the standard deviation of the population of errors is the standard deviation of the sample of estimated errors, that is, of the residuals $\hat{e}_i = y_i - \hat{\alpha} - \hat{\beta}x_i$. This estimate is $\sqrt{\Sigma \hat{e}_i^2/(n-1)}$, but instead of using this, we modify it slightly and use formula (3.3). The modification has the nice feature that $\hat{\sigma}^2$ is an unbiased estimate of σ^2.

The divisor $n - 2$ in formula (3.3) is sometimes called the *degrees of freedom* of the estimate $\hat{\sigma}^2$. The subtraction of 2 from n corresponds to the fact that we must estimate two parameters α and β in order to form the residuals $\hat{e}_i$.

The p-Value. We argued earlier in this section that a very large value of $|t|$ is strong evidence against $\beta = 0$ because such a value of $|t|$ would be very unlikely if $\beta = 0$. In other words, we can measure the strength of the observed data's evidence against $\beta = 0$ by how unlikely the observed $|t|$ would be if $\beta = 0$. This is the idea behind p-values.

To define the p-value of the test, we must distinguish between two different views of $|t|$ in (3.5). The first view is to view $|t|$ as the observed value of $|t|$, that is, the value obtained by substituting the observed values of the y_i's into the formula. The second view is to view the y_i's and hence $|t|$ as random variables. The *p-value* of the test is the probability, assuming $\beta = 0$, that the random variable $|t|$ is as large or larger than the observed $|t|$. To interpret this, suppose model (3.1) is true with $\beta = 0$ and imagine the acid content experiment is repeated an infinite number of times. For each experiment, the test statistic $|t|$ would be calculated. The p-value is the proportion of this imagined infinite collection of $|t|$'s that would be as large or larger than the value of $|t|$ calculated from the actual experiment.

If the p-value is very small, then it is very unlikely that $\beta = 0$ and so we conclude $\beta \neq 0$. Otherwise, we conclude $\beta = 0$. It can be left to the judgement of the individual to say when a p-value is "very small". Most statisticians would consider 0.01 to be very small and not many would consider 0.10 to be very small. For p-values between 0.01 and 0.10 judgements vary.

Note the difference in our confidence about the two conclusions $\beta \neq 0$ and $\beta = 0$. When the p-value is very small and we conclude $\beta \neq 0$, we are quite sure about our conclusion because the evidence is strong. But when we conclude $\beta = 0$, we are not convinced that $\beta = 0$ exactly but are only concluding that $\beta = 0$ is plausible and provides an adequate model for the data.

The Distribution of t. In order to calculate the p-value we must know the probability distribution of the random variable $|t|$, or of t, when $\beta = 0$. The test statistic t is obtained from the data, that is, from the x_i's and y_i's. The x_i's are assumed to be nonrandom constants, so the randomness of t derives from the randomness of the y_i's. Note that the probability distribution of y_i is not completely known, even if we suppose $\beta = 0$, because then $y_i = \alpha + e_i$ and α is an unknown parameter and the distribution of e_i is not completely known; in particular, the standard deviation σ of e_i cannot be assumed to be known. However, by inspecting the formula for t it can be seen that t depends on the y_i's only through the difference $y_i - \bar{y}$, and the parameter α cancels in these differences. Moreover, the parameter σ cancels in the ratio of $\hat{\beta}$ to est.SD($\hat{\beta}$). Therefore, if we specify the shape of the distribution of the random errors, leaving only the standard deviation unknown, then the distribution of t is completely known.

If we specify the shape to be normal (bell-shaped), that is, if we assume the normal linear regression model, then the resulting distribution of t, when $\beta = 0$, is called the *t distribution* with $n - 2$ degrees of freedom. (The degrees of freedom are associated with the estimate $\hat{\sigma}$ of σ.) Even if the distribution of the random errors is not normal, the distribution of the test statistic t, when $\beta = 0$, is still close to the t distribution with $n - 2$ degrees of freedom, provided that the sample size n is large. It should be safe to rely on the t distribution with a sample size as large as 20, provided the normality of the distribution of the random errors has been checked (see the next section) and the data, if necessary, have been transformed.

The Acid Content Data. For the acid content data, the degrees of freedom are 18 (= 20 − 2). Above we calculated $|t|$ to be about 57. We can use the t table in the Appendix to pin down the p-value. Looking in the table in the row labeled 18, we see the numbers 1.33 1.73 2.10 2.88 3.92. Locate 57 relative to these numbers; it is greater than the last entry 3.92. This entry is in the column labeled 0.001. This means that the probability, when $\beta = 0$, that the random variable $|t|$ is as large or larger than 3.92 is 0.001, or in compact notation,

$$\text{Prob}[|t| \geq 3.92] = 0.001$$

Of course the probability that $|t|$ is as large or larger than 57 is even less. So we can say that the p-value of the test of $\beta = 0$ for the acid content data is less than 0.001. This is a formal confirmation of our previous informal conclusion that $\beta \neq 0$. In other words, as we would expect, the cheap measurement of the acid number of a chemical sample contains significant information about the organic acid content of the sample.

3.5 CHECKING NORMALITY

Least-squares tests and estimates are optimal if the population of errors can be assumed to have a normal distribution. If the normality assumption is not satisfied, then least-squares procedures are still valid but they may be far from optimal. Consider the least-squares test of $\beta = 0$ described in the preceding section and suppose the error population is nonnormal. The least-squares test is still approximately valid (provided the sample size is not too small and the nonnormality is not too extreme) in the sense that the calculated p-value is approximately equal to the true p-value, but there are other tests that are more powerful in the sense that they are better at detecting when $\beta \neq 0$.

There are two approaches to dealing with the question of normality. One way is to check the data for normality and, if nonnormality is detected, try to correct it. Or, one can use regression methods other than least-squares, such as those presented in Chapters 4, 5, and 6, which do not depend on the assumption of normality.

A number of plots and tests, based on the residuals, have been developed for checking the normality of the errors, but here we mention only the normal probability plot. The standardized residuals are put in increasing order and are plotted against what their expected values would be if they came from a sample of n independent standard normal random variables. The plot should look nearly linear if the assumption of normality is valid. A normal probability plot of the residuals from the acid content data, shown in Figure 3.4, looks sufficiently linear to be consistent with the assumption of normality.

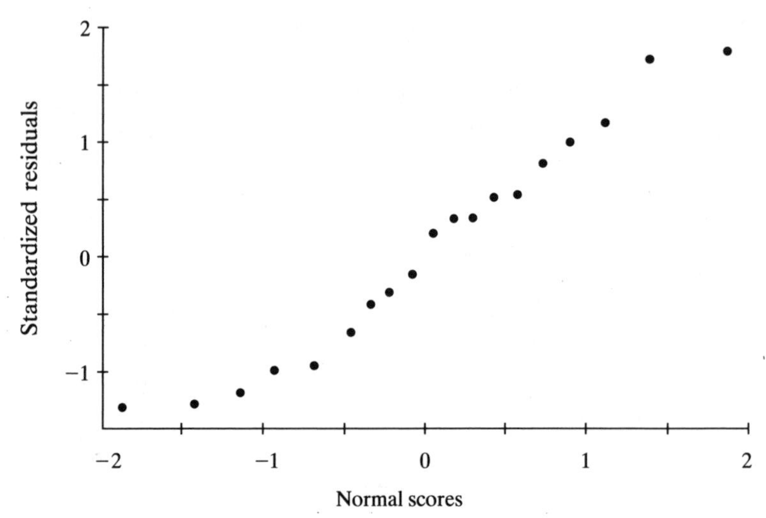

Figure 3.4 Normal probability plot of the residuals for the acid content data.

When a normal probability plot is very nonlinear, the data can sometimes be transformed so that normality is more closely approximated.

3.6 AN EXAMPLE OF MULTIPLE REGRESSION

Simple regression is when there is only one explanatory variable and *multiple* regression is when there are more than one. The turnip green data set in Section 1.2 has three explanatory variables. Let us do a regression analysis of these data. The linear regression model for these data is

$$Y = \beta_0 + \beta_1 X_1 + \beta_2 X_2 + \beta_3 X_3 + e$$

In terms of the observed data the model is

$$y_i = \beta_0 + \beta_1 x_{i1} + \beta_2 x_{i2} + \beta_3 x_{i3} + e_i \tag{3.6}$$

for $i = 1, 2, \ldots, 27$.

3.7 ESTIMATING THE REGRESSION COEFFICIENTS

The discussion in Section 3.3 extends to multiple regression. The *least-squares estimates* of β_0, β_1, β_2, and β_3 are defined to be the values of $\hat{\beta}_0$, $\hat{\beta}_1$, $\hat{\beta}_2$, and $\hat{\beta}_3$ which give the least sum of squares of the residuals, that is, which give the least value of $\Sigma\hat{e}_i^2$, where $\hat{e}_i = y_i - (\hat{\beta}_0 + \hat{\beta}_1 x_{i1} + \hat{\beta}_2 x_{i2} + \hat{\beta}_3 x_{i3})$.

Algebraic formulas for these four least-squares estimates are very messy if they are written out for each individual parameter. But if we introduce matrix notation the formula for the vector of estimates is quite compact.

Matrix Notation. Boldface letters are used to denote matrices and vectors, capital letters for matrices and lowercase letters for vectors. Vectors in formulas are taken to be column vectors. (Elsewhere, it is sometimes more convenient to write a vector as a row vector.) Let

$$\mathbf{y} = \begin{bmatrix} y_1 \\ y_2 \\ \vdots \\ y_n \end{bmatrix}, \quad \mathbf{X} = \begin{bmatrix} 1 & x_{11} & x_{12} & x_{13} \\ 1 & x_{21} & x_{22} & x_{23} \\ \vdots & \vdots & \vdots & \vdots \\ 1 & x_{n1} & x_{n2} & x_{n3} \end{bmatrix}, \quad \boldsymbol{\beta} = \begin{bmatrix} \beta_0 \\ \beta_1 \\ \beta_2 \\ \beta_3 \end{bmatrix}, \quad \mathbf{e} = \begin{bmatrix} e_1 \\ e_2 \\ \vdots \\ e_n \end{bmatrix}$$

where $n = 27$. Model (3.6) can be reexpressed in matrix notation as

$$\mathbf{y} = \mathbf{X}\boldsymbol{\beta} + \mathbf{e} \tag{3.7}$$

The formula for the vector of least-squares regression estimates is

$$\hat{\boldsymbol{\beta}} = (\mathbf{X}'\mathbf{X})^{-1}\mathbf{X}'\mathbf{y} \tag{3.8}$$

The General Model. Consider a data set with n observations and p explanatory variables, as shown in Table 1.2. Let $\mathbf{y}$ denote the column labeled Y and let $\mathbf{X}$ denote the p columns labeled $X_1, \ldots, X_p$ with a column of 1's put in front. Let

$$\boldsymbol{\beta} = \begin{bmatrix} \beta_0 \\ \beta_1 \\ \vdots \\ \beta_p \end{bmatrix}, \quad \mathbf{e} = \begin{bmatrix} e_1 \\ e_2 \\ \vdots \\ e_n \end{bmatrix}$$

The linear regression model for the data in Table 1.2 is shown in equation (1.2). The matrix notation for the model and for the vector of least-squares regression estimates are exactly equations (3.7) and (3.8) above.

The Case of Simple Regression. Simple regression corresponds to $p = 1$. So if we let $p = 1$ in formula (3.8) we should get the same estimates as in formulas (3.2). This is not obvious just by looking at the formulas. As a check, let us apply (3.8) to the acid content data in Table 3.1 to see whether the estimates coincide with those obtained in Section 3.3.

To use formula (3.8) we form

$$y = \begin{bmatrix} 76 \\ 70 \\ 55 \\ 71 \\ 55 \\ 48 \\ 50 \\ 66 \\ 41 \\ 43 \\ 82 \\ 68 \\ 88 \\ 58 \\ 64 \\ 88 \\ 89 \\ 88 \\ 84 \\ 88 \end{bmatrix}, \qquad X = \begin{bmatrix} 1 & 123 \\ 1 & 109 \\ 1 & 62 \\ 1 & 104 \\ 1 & 57 \\ 1 & 37 \\ 1 & 44 \\ 1 & 100 \\ 1 & 16 \\ 1 & 28 \\ 1 & 138 \\ 1 & 105 \\ 1 & 159 \\ 1 & 75 \\ 1 & 88 \\ 1 & 164 \\ 1 & 169 \\ 1 & 167 \\ 1 & 149 \\ 1 & 167 \end{bmatrix}$$

and we calculate

$$X'y = \begin{bmatrix} 1372 \\ 157154 \end{bmatrix}, \qquad X'X = \begin{bmatrix} 20 & 2061 \\ 2061 & 261419 \end{bmatrix}$$

$$(X'X)^{-1} = \begin{bmatrix} 0.266575 & -0.00210165 \\ -0.00210165 & 0.0000203944 \end{bmatrix}$$

$$\hat{\beta} = \begin{bmatrix} 0.266575 & -0.00210165 \\ -0.00210165 & 0.0000203944 \end{bmatrix} \begin{bmatrix} 1372 \\ 157154 \end{bmatrix} = \begin{bmatrix} 35.46 \\ 0.3216 \end{bmatrix}$$

Therefore the estimated regression line is $\hat{Y} = 35.46 + 0.3216X$, the same as in Section 3.3.

Geometric Interpretation of Formula (3.8). It is not yet clear what formula (3.8) has to do with minimizing the sum of squares of the residuals. To make the connection, let us take a geometric view. First note that the vector of residuals is $\hat{e} = y - \hat{y}$, where the vector $\hat{y}$ of fitted y-values comes from the estimated regression equation $\hat{y} = X\hat{\beta}$. It is a vector in n-dimensional space but in order to be able to visualize the geometry, think of the case $n = 3$, so that $\hat{e}$ is a vector in ordinary three-dimensional space. The length of $\hat{e}$ is $\sqrt{\Sigma \hat{e}_i^2}$, so minimizing the sum of squares of the residuals is the same as minimizing the length of $\hat{e}$, which is the distance between y and $\hat{y}$.

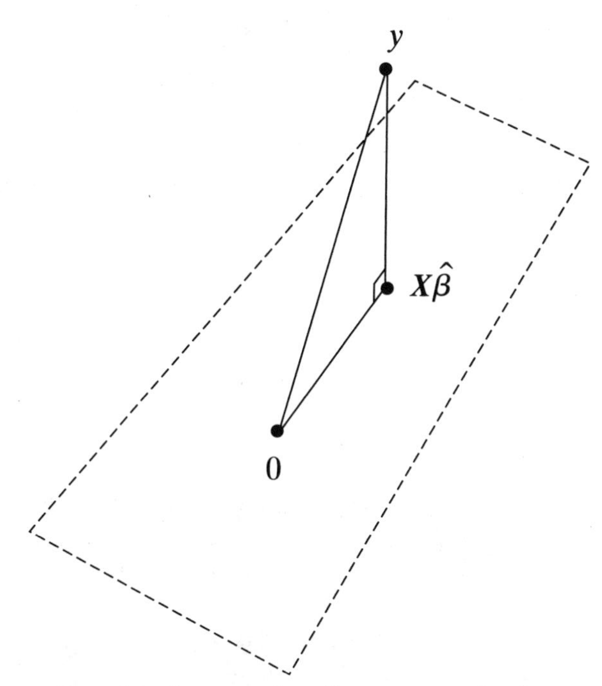

Figure 3.5 Geometric view of least-squares estimation.

Among all candidates $\boldsymbol{b}$, the distance between $\boldsymbol{y}$ and $\boldsymbol{Xb}$ is minimized by the choice $\boldsymbol{b} = \hat{\boldsymbol{\beta}}$. Think of the set of all $\boldsymbol{Xb}$ as a plane cutting through three-dimensional space. See Figure 3.5. We want to find the point in this plane that is closest to the point $\boldsymbol{y}$. Intuition tells us that the shortest distance from the point to the plane is achieved along a line *perpendicular* to the plane. Therefore $\hat{\boldsymbol{\beta}}$ is characterized by the fact that the line from $\boldsymbol{y}$ to $\boldsymbol{X}\hat{\boldsymbol{\beta}}$, that is, the vector $\boldsymbol{X}\hat{\boldsymbol{\beta}} - \boldsymbol{y}$, is perpendicular to the plane of all $\boldsymbol{Xb}$. This can be expressed as $\boldsymbol{X}'(\boldsymbol{X}\hat{\boldsymbol{\beta}} - \boldsymbol{y}) = \boldsymbol{0}$, or $\boldsymbol{X}'\boldsymbol{X}\hat{\boldsymbol{\beta}} = \boldsymbol{X}'\boldsymbol{y}$, which is equivalent to formula (3.8).

Algebraic Justification of Formula (3.8). Some readers may be more convinced by an algebraic argument than by a geometric argument. For a candidate $\boldsymbol{b}$ for the vector of least-squares estimates, the sum of squares of the residuals is $(\boldsymbol{y} - \boldsymbol{Xb})'(\boldsymbol{y} - \boldsymbol{Xb})$, which can be written as $\boldsymbol{y}'\boldsymbol{y} - 2\boldsymbol{y}'\boldsymbol{Xb} + \boldsymbol{b}'\boldsymbol{X}'\boldsymbol{Xb}$. Considering this as a function of the components $b_0, b_1, \ldots, b_p$ of $\boldsymbol{b}$, its minimum can be found by taking partial derivatives with respect to the b_j's and setting them equal to 0. It is convenient to take partial derivatives using matrix notation. First consider the special case $p = 0$. Then the function would be $y^2 - 2yxb + x^2b^2$ and the derivative with respect to b would be

$-2yx + 2x^2b$. For general p the vector of partial derivatives is analogous: $-2\mathbf{X}'\mathbf{y} + 2\mathbf{X}'\mathbf{X}\mathbf{b}$. Setting this equal to $\mathbf{0}$, we get $\mathbf{X}'\mathbf{X}\mathbf{b} = \mathbf{X}'\mathbf{y}$, which coincides with (3.8).

The Turnip Green Data. For the data in Table 1.1 formula (3.8) yields the vector of estimates

$$\begin{bmatrix} \hat{\beta}_0 \\ \hat{\beta}_1 \\ \hat{\beta}_2 \\ \hat{\beta}_3 \end{bmatrix} = \begin{bmatrix} 82.07 \\ 0.02276 \\ -0.7783 \\ 0.1640 \end{bmatrix}$$

So the estimated regression equation is $\hat{Y} = 82.07 + 0.02276X_1 - 0.7783X_2 + 0.1640X_3$. (If you want to verify this, use a computer. Calculation of formula (3.8) using an ordinary hand calculator is feasible for $p = 1$ but is difficult for $p = 3$.)

A more complete analysis of these data is outlined in Example 1 in Section 2.4. At the beginning of the analysis, model (3.6) is used, but it is found that the following model is better:

$$Y = \beta_0 + \beta_1 X_1 + \beta_2 X_2 + \beta_3 X_3 + \beta_4 X_4 + e \tag{3.9}$$

where $X_4 = X_2^2$. The estimated regression equation is $\hat{Y} = 119.6 - 0.03367X_1 + 5.425X_2 - 0.5026X_3 - 0.1209X_2^2$.

3.8 TESTING THE REGRESSION COEFFICIENTS

The first test performed in a regression analysis is often a test of whether the explanatory variables actually contain any significant explanatory information. Let us perform such a test for the turnip green data. We want to compare the full model (3.9), containing all four explanatory variables, with the reduced model $Y = \beta_0 + e$, containing no explanatory variables, to see whether there is a significant difference between these two models. In other words, we want to test $\beta_1 = \beta_2 = \beta_3 = \beta_4 = 0$.

In developing a test of $\beta = 0$ in Section 3.4 we started by noting that the value of $\hat{\beta}$ should tell us whether or not $\beta = 0$. Similarly, it makes sense that the values of $\hat{\beta}_1$, $\hat{\beta}_2$, $\hat{\beta}_3$, and $\hat{\beta}_4$ should tell us whether or not $\beta_1 = \beta_2 = \beta_3 = \beta_4 = 0$. But rather than develop a test from the viewpoint of testing whether certain parameters are zero, we take the alternative viewpoint of comparing two models.

A Test Statistic. The suitability of a model can be judged by the size of the residuals. The smaller the residuals, the better the model fits the data. In the least-squares method, an overall measure of the size of the residuals is given by the sum of squares of the residuals. Let SSR denote the sum of squares of the residuals of a model. We can compare the full model with the reduced model by comparing SSR_{full} with $\text{SSR}_{\text{reduced}}$. Specifically, the test statistic we use for testing $\beta_1 = \beta_2 = \beta_3 = \beta_4 = 0$ is

$$F = \frac{\text{SSR}_{\text{reduced}} - \text{SSR}_{\text{full}}}{4\hat{\sigma}^2} \tag{3.10}$$

where $\hat{\sigma}^2$ is an estimate of the variance σ^2 of the distribution of the random errors.

Estimating $\boldsymbol{\sigma^2}$. A natural estimate of the variance of the population of errors is the variance of the sample of estimated errors, that is, the residuals $\hat{e}_i = y_i - (\hat{\beta}_0 + \hat{\beta}_1 x_{i1} + \hat{\beta}_2 x_{i2} + \hat{\beta}_3 x_{i3} + \hat{\beta}_4 x_{i4})$. In order to make $\hat{\sigma}^2$ an unbiased estimate of σ^2 we define

$$\hat{\sigma}^2 = \frac{\Sigma \hat{e}_i^2}{n - 5} \tag{3.11}$$

where the divisor $n - 5 = 22$ is used rather than $n - 1 = 26$. The subtraction of 5 from n corresponds to the fact that we must estimate five parameters β_0, β_1, β_2, β_3, and β_4 in order to form the residuals $\hat{e}_i$. Note that $\Sigma \hat{e}_i^2$ is the same as SSR_{full}.

Justification of Formula (3.10). The reduced model cannot possibly fit the data as well as the full model because it has fewer parameters and hence is less flexible. So it is always true that $\text{SSR}_{\text{reduced}}$ is larger than SSR_{full} and the difference $\text{SSR}_{\text{reduced}} - \text{SSR}_{\text{full}}$ is positive. But when the reduced model is true, we expect this difference to be smaller than when the reduced model is false. It can be shown that when the reduced model is true, then the expected value of $\text{SSR}_{\text{reduced}} - \text{SSR}_{\text{full}}$ is $4\sigma^2$. (The multiplier 4 in $4\sigma^2$ is the same as the number of parameters set equal to 0 in the hypothesis $\beta_1 = \beta_2 = \beta_3 = \beta_4 = 0$.) So when $\beta_1 = \beta_2 = \beta_3 = \beta_4 = 0$, we expect F to be close to 1. Having a preference for simpler models, we will decide $\beta_1 = \beta_2 = \beta_3 = \beta_4 = 0$ unless there is strong evidence against it as shown by a value of F that is much larger than 1.

The *p*-Value. The strength of the evidence against the null hypothesis $\beta_1 = \beta_2 = \beta_3 = \beta_4 = 0$ is measured by the largeness of F, which in turn is measured by the smallness of the p-value. The p-value of the test is the probability, assuming the null hypothesis, that the random variable F is as large or larger than the observed value of F. To interpret the meaning of the p-value, we should imagine repeating the turnip green experiment, that is, picking leaves from another random sample of 27 turnip plants and measuring the vitamin B_2 concentration, sunlight, soil moisture, and air temperature for each plant. Imagine repeating this experiment an infinite number of times. For each experiment, calculate the value of the test statistic F. If the null hypothesis is true, then the p-value is the proportion of these values of F which would be as large or larger than the value of F in the original experiment.

Suppose the p-value of a test is very small, say 0.01. This says that the observed value of F is so large that it would be very improbable if the null hypothesis were true. Rather than believe that such an unlikely event has occurred, it is more sensible to conclude that the null hypothesis is false.

To calculate the p-value we need to know the distribution of the random variable F when the null hypothesis is true.

The Distribution of F. The distribution of F depends on the shape of the distribution of the random errors. For testing hypotheses by the method of least-squares, we assume that the errors are normally distributed. The resulting distribution of F, when the null hypothesis is true, is the F *distribution* with 4 and $n - 5$ degrees of freedom.

The Turnip Green Data. From the data in Table 1.1 we can calculate $\text{SSR}_{\text{full}} = 819.7$, $\hat{\sigma}^2 = 819.7/22 = 37.26$, and $\text{SSR}_{\text{reduced}} = 9150.5$. So $F = (9150.5 - 819.7)/4(37.26) = 55.90$. The degrees of freedom are 4 and 22.

The calculation of SSR_{full} is very difficult to do with an ordinary hand calculator, and so one would want to have the calculation of the test statistic done by a regression computer package or else write a computer program oneself (see Section 3.12). Most regression computer packages would print out the p-value of the test. Some computer languages include a function that would yield the p-value given the input 55.90, 4, and 22. The p-value is less than 0.0001.

Without a computer, one can pin down the p-value by using tables called F tables. The F table in the Appendix tells us that for 4 and 22 degrees of freedom, $\text{Prob}[F \geq 4.31] = 0.01$ when the null hypothesis is true. Since $55.90 > 4.31$, we see that the p-value of our test is less than 0.01. So we can conclude that vitamin B_2 concentration in turnip greens has a significant linear relationship with at least some of the four variables sunlight, soil moisture, air temperature, and squared soil moisture.

3.9 TESTING $\beta_{q+1} = \cdots = \beta_p = 0$

Consider the general form of the linear regression model, $Y = \beta_0 + \beta_1 X_1 + \cdots + \beta_p X_p + e$. The testing procedure described in the preceding section can be extended to allow comparison of this model with the reduced model formed from any subset of the explanatory variables. By changing indices, we can write the reduced model as $Y = \beta_0 + \beta_1 X_1 + \cdots + \beta_q X_q$, where $q < p$. Note that comparing these two models is the same as testing $\beta_{q+1} = \cdots = \beta_p = 0$.

We continue to use the notation SSR_{full} and $\text{SSR}_{\text{reduced}}$ for the sums of squares of the residuals for the two models. Let $\hat{\sigma}^2$ denote the following unbiased estimate of σ^2:

$$\hat{\sigma}^2 = \frac{\text{SSR}_{\text{full}}}{n - p - 1} \tag{3.12}$$

The test statistic for testing the null hypothesis $\beta_{q+1} = \cdots = \beta_p = 0$ is

$$F = \frac{\text{SSR}_{\text{reduced}} - \text{SSR}_{\text{full}}}{(p - q)\hat{\sigma}^2} \tag{3.13}$$

Assuming normality of the distribution of random errors, F has an F distribution with $p - q$ and $n - p - 1$ degrees of freedom when the null hypothesis is true.

3.10 TESTING $\beta_3 = 0$

The test in Section 3.8 indicates that the variables of sunlight, soil moisture, and air temperature contain significant explanatory information about vitamin B_2 concentration in turnip greens. But maybe most of this information is contained in just one or two of the three variables. A test of $\beta_3 = 0$ is a test of whether most of the information is in sunlight and soil moisture (and its square) so that, given this information, only an insignificant amount of additional information is in the air temperature.

A Test Based on a Comparison of Models. Testing $\beta_3 = 0$ is, if we interchange the indices 3 and 4, a special case of testing $\beta_{q+1} = \cdots = \beta_p = 0$ in which $p = 4$ and $q = 3$. Therefore, according to Section 3.9, a reasonable test statistic is

$$F = \frac{\mathrm{SSR}_{\mathrm{reduced}} - \mathrm{SSR}_{\mathrm{full}}}{\hat{\sigma}^2}$$

where $\hat{\sigma}^2 = \mathrm{SSR}_{\mathrm{full}}/(n-5)$. In terms of the original indices, the full model is $Y = \beta_0 + \beta_1 X_1 + \beta_2 X_2 + \beta_3 X_3 + \beta_4 X_2^2 + e$ and the reduced model is $Y = \beta_0 + \beta_1 X_1 + \beta_2 X_2 + \beta_4 X_2^2 + e$. Assuming normality of the distribution of random errors, F has an F distribution with 1 and $n-5$ degrees of freedom when $\beta_3 = 0$.

The value of F for testing $\beta_3 = 0$ in the turnip green data is 5.68 and the p-value is 0.03. Since the p-value is less than 0.05, most statisticians would reject the null hypothesis and conclude that the air temperature variable makes a significant contribution to the model.

A Test Based on $\hat{\beta}_3$. The value of $\hat{\beta}_3$ should indicate whether $\beta_3 = 0$ or not. Similar to the test statistic (3.5) for testing $\beta = 0$ in the case of simple regression,

$$|t| = \frac{|\hat{\beta}_3|}{\mathrm{est.SD}(\hat{\beta}_3)}$$

is a reasonable test statistic for testing $\beta_3 = 0$. Now we need to know how to obtain $\mathrm{est.SD}(\hat{\beta}_3)$.

In simple least-squares regression, $\mathrm{Var}(\hat{\beta}) = \sigma^2/\Sigma(x_i - \bar{x})^2$. For the acid content data, this formula yields $\mathrm{Var}(\hat{\beta}) = 0.0000203944\sigma^2$. Note that 0.0000203944 appears in the lower right corner of the matrix $(X'X)^{-1}$ in the subsection on simple regression in Section 3.7. In fact, in simple regression, $\mathrm{Var}(\hat{\beta})$ can always be obtained by multiplying σ^2 by the second diagonal entry in $(X'X)^{-1}$. Similarly, in multiple regression with four explanatory variables, $\mathrm{Var}(\hat{\beta}_3)$ can be obtained by multiplying σ^2 by the fourth diagonal entry in $(X'X)^{-1}$. In the general linear regression model, $\mathrm{Var}(\hat{\beta}_j)$ is equal to σ^2 times the $(j+1)$th diagonal entry in $(X'X)^{-1}$ for any $j = 0, 1, \ldots, p$. Estimating σ^2 by $\hat{\sigma}^2 = \mathrm{SSR}_{\mathrm{full}}/(n-5)$, we obtain $\mathrm{est.Var}(\hat{\beta}_3)$, and taking the square root, we obtain $\mathrm{est.SD}(\hat{\beta}_3)$.

When $\beta_3 = 0$, t has a t distribution with $n-5$ degrees of freedom, which are the degrees of freedom associated with $\hat{\sigma}^2$.

For the turnip green data, $\hat{\beta}_3 = -0.5026$ and $\mathrm{est.SD}(\hat{\beta}_3) = 0.2109$, so $|t| = 0.5026/0.2109 = 2.38$. The p-value is 0.03.

Note that the p-values are the same for the F test based on a comparison of models and for the t test based on $\hat{\beta}_3$. In fact, the two tests are equivalent and their test statistics are related by the equation $F = t^2$. For the turnip green data, $F = 5.68$ and $t^2 = (2.38)^2 = 5.66$. (The difference between 5.68 and 5.66 is due to round-off error.)

3.11 THE COEFFICIENT OF DETERMINATION

The *coefficient of determination* is a measure of how well the explanatory variables explain the response variable. It is defined to be

$$R^2 = \frac{\Sigma(\hat{y}_i - \bar{y})^2}{\Sigma(y_i - \bar{y})^2}$$

An observed value y_i of the response variable differs from the average by the deviation $y_i - \bar{y}$. The deviation can be split into two parts: $y_i - \bar{y} = (y_i - \hat{y}_i) + (\hat{y}_i - \bar{y})$. The second part is the part of the deviation that is explained by the relationship $\hat{y}_i = \hat{\beta}_0 + \hat{\beta}_1 x_{i1} + \cdots + \hat{\beta}_p x_{ip}$ of the response variable with the explanatory variables. The sum $\Sigma(y_i - \bar{y})^2$ can be regarded as the total variability observed in the response variable, and the sum $\Sigma(\hat{y}_i - \bar{y})^2$ can be regarded as the amount of this variability that is explained. Thus R^2 is the proportion of the total variability in the response variable that is explained by the explanatory variables.

The coefficient of determination is closely related to the test statistic F for testing $\beta_1 = \beta_2 = \cdots = \beta_p = 0$. In fact, R^2 can be expressed as a function of F.

In the case of simple regression, R^2 is equal to the square of the sample correlation between X and Y.

3.12 COMPUTATION

Use of a Hand Calculator for Simple Regression. Some hand calculators have special keys for doing simple linear regression. Then you only need to enter the numbers $y_1, x_1, y_2, x_2, \ldots, y_n, x_n$ and push a few special keys to obtain the least-squares regression estimates $\hat{\alpha}$ and $\hat{\beta}$. On a hand calculator without such capability, you could organize the calculations as follows. First calculate $S_x = \Sigma x_i$, $S_y = \Sigma y_i$, $S_{xx} = \Sigma x_i^2$, $S_{yy} = \Sigma y_i^2$, and $S_{xy} = \Sigma x_i y_i$. Then $\hat{\beta} = (nS_{xy} - S_x S_y)/(nS_{xx} - S_x^2)$ and $\hat{\alpha} = (S_y - \hat{\beta} S_x)/n$. You should retain a large number of digits in these calculations even if you only desire accuracy in $\hat{\alpha}$ and $\hat{\beta}$ to a few significant digits. This is because the subtractions in this formula for $\hat{\beta}$ often involve numbers that are nearly equal. Of course formulas (3.2) could be used to calculate $\hat{\alpha}$ and $\hat{\beta}$, but it is inconvenient to deal with the deviations $x_i - \bar{x}$ and $y_i - \bar{y}$ on a hand calculator.

You can also use a hand calculator to obtain the test statistic $|t|$ for testing $\beta = 0$. Calculate $\hat{\sigma}$ as the square root of $[(nS_{yy} - S_y^2) - \hat{\beta}^2(nS_{xx} - S_x^2)]/n(n-2)$. This is more convenient than dealing with the residuals $\hat{e}_i = y_i - (\hat{\alpha} + \hat{\beta}x_i)$ in formula (3.3). Divide $\hat{\sigma}$ by the square root of $(nS_{xx} - S_x^2)/n$ to obtain est.SD$(\hat{\beta})$. Then divide $|\hat{\beta}|$ by est.SD$(\hat{\beta})$ to obtain $|t|$. The p-value can be determined approximately from the t table in the Appendix. For example, suppose $n = 11$ and you calculate $|t| = 2.03$. Look in the table in the row labeled 9 ($= 11 - 2$). Note that 2.03 falls between the numbers 1.83 and 2.26 in the columns labeled 0.10 and 0.05. Therefore the p-value of the test falls between 0.05 and 0.10.

Use of a Computer. Many computer packages are available for performing least-squares regression, such as BMDP, Minitab, SAS, and SPSS. Or, in a computer language such as APL, GAUSS, or SAS/IML, in which matrix manipulation is straightforward, it is not hard to write your own program for least-squares regression, using formulas (3.8) and (3.13). For calculating SSR_{full}, note that $\hat{\boldsymbol{e}} = \boldsymbol{y} - \boldsymbol{X}\hat{\boldsymbol{\beta}}$ and $\text{SSR}_{\text{full}} = \hat{\boldsymbol{e}}'\hat{\boldsymbol{e}}$. This also works for calculating $\text{SSR}_{\text{reduced}}$ if $\boldsymbol{X}$ is reduced to $\boldsymbol{X}_{\text{reduced}}$ by omitting the columns corresponding to the coefficients set equal to 0 in the null hypothesis. New vectors $\hat{\boldsymbol{\beta}}$ and $\hat{\boldsymbol{e}}$ are calculated based on $\boldsymbol{X}_{\text{reduced}}$ and then $\text{SSR}_{\text{reduced}} = \hat{\boldsymbol{e}}'\hat{\boldsymbol{e}}$. In the case $q = 0$, $\text{SSR}_{\text{reduced}}$ is simply $\Sigma(y_i - \bar{y})^2$.

Test Case. To try out a computational procedure for least-squares regression on your calculator or computer, the data set in Table 3.2 can serve as a quick test case.

For simple regression, let $X = X_1$. The estimated regression line is $\hat{Y} = 33.16 + 1.480X$. The estimate of σ is $\hat{\sigma} = 3.552$ and the test statistic for testing $\beta = 0$ is $|t| = 2.946$.

For multiple regression, using both X_1 and X_2 as explanatory variables, the estimated regression equation is $\hat{Y} = 51.30 + 0.6025X_1 - 0.6180X_2$. The estimate of σ is $\hat{\sigma} = 3.328$ and the test statistic for testing $\beta_1 = \beta_2 = 0$ is $F = 5.722$.

Table 3.2 Test Case for Linear Regression

Y	X_1	X_2
37	4	22
40	6	24
48	6	18
44	9	20
50	11	15
51	12	9

NOTES

3.1. Legendre (1752–1833) and Gauss (1777–1855) both made many important contributions to pure and applied mathematics. Gauss is generally regarded as one of the greatest mathematicians of all time. Gauss did not publish anything on least squares until 1809, but he claimed to have been using the method since 1795, which led to a dispute over who deserved credit for its discovery. Stigler (1986) finds Gauss's claim of prior discovery likely to be true but he credits Legendre with independent discovery and with prior recognition of the importance of least squares as a general method for combining inexact measurements.

3.3a. The details of the derivation of (3.2) are as follows. The partial derivative of $\Sigma(y_i - a - bx_i)^2$ with respect to a is equal to 0 when $\Sigma(y_i - a - bx_i) = 0$, or $\bar{y} - a - b\bar{x} = 0$, or $a = \bar{y} - b\bar{x}$, which coincides with the second formula in (3.2). The partial derivative with respect to b is equal to 0 when $\Sigma x_i(y_i - a - bx_i) = 0$. Substitute $a = \bar{y} - b\bar{x}$ into this equation to obtain $\Sigma x_i[(y_i - \bar{y}) - b(x_i - \bar{x})] = 0$, or $b = \Sigma x_i(y_i - \bar{y})/\Sigma x_i(x_i - \bar{x})$. This coincides with the first formula in (3.2) because $\Sigma x_i(y_i - \bar{y}) = \Sigma(x_i - \bar{x})(y_i - \bar{y})$ and $\Sigma x_i(x_i - \bar{x}) = \Sigma(x_i - \bar{x})^2$, because $\Sigma(y_i - \bar{y}) = 0$ and $\Sigma(x_i - \bar{x}) = 0$.

3.3b. In the expression $w_i = (x_i - \bar{x})^2/\Sigma(x_i - \bar{x})^2$, the subscript i plays two different roles. The subscript i in w_i matches only with the i in the numerator of the ratio. The i in the denominator matches with a suppressed i in the summation notation. More precisely (but also more cumbersomely) we can write $w_i = (x_i - \bar{x})^2/\Sigma_{i=1}^n(x_i - \bar{x})^2$, or even less ambiguously, $w_i = (x_i - \bar{x})^2/\Sigma_{k=1}^n(x_k - \bar{x})^2$.

3.3c. Another good feature of the least-squares estimates $\hat{\alpha}$ and $\hat{\beta}$ is that they are *unbiased* estimates. That is, when viewed as random variables, their expectations are α and β, respectively.

3.3d. For completeness, we include here a review of expectation. The *expectation* of the random variable $\hat{\beta}$, denoted $E(\hat{\beta})$, can be informally defined as follows.

First, a helpful way to think of $\hat{\beta}$ as a random variable is to imagine repetitions of the experiment. Imagine repeating the acid content experiment by taking 20 more samples of the chemical and using both the cheap and expensive measurement procedures on each sample. From these new data we would calculate $\hat{\beta}$. Imagine repeating the experiment an infinite number of times. From the data of each experiment we would calculate $\hat{\beta}$. It can be helpful to interpret the properties of the random variable $\hat{\beta}$ in terms of this imagined infinite collection of $\hat{\beta}$'s.

The expectation of the random variable $\hat{\beta}$ is the average of this infinite collection of $\hat{\beta}$'s. It can be regarded as being at the "center" of the imagined collection of $\hat{\beta}$'s. (But other definitions of "center" are also possible.) It can be shown that if model (3.1) is true, then $E(\hat{\beta}) = \beta$.

3.4a. To derive (3.4), note that $\hat{\beta} = \Sigma c_i y_i$, where $c_i = (x_i - \bar{x})/\Sigma(x_i - \bar{x})^2$. This uses the fact that $\Sigma(x_i - \bar{x})(y_i - \bar{y}) = \Sigma(x_i - \bar{x})y_i$, which is true because $\Sigma(x_i - \bar{x}) = 0$. The x_i's are considered to be nonrandom constants. (Even if the x_i's were obtained randomly, it is still possible to regard them as constants through the concept of conditional distributions.) Given that the x_i's are nonrandom, the only random part of $y_i = \alpha + \beta x_i + e_i$ is e_i. Hence $\text{Var}(y_i) = \text{Var}(e_i) = \sigma^2$ and the y_i's are independent of one another. Then

$$\begin{aligned}\text{Var}(\hat{\beta}) &= \text{Var}\left(\sum c_i y_i\right) = \sum c_i^2 \,\text{Var}(y_i) = \left(\sum c_i^2\right)\sigma^2 \\ &= \left[1\Big/\sum (x_i - \bar{x})^2\right]\sigma^2 = \sigma^2\Big/\sum (x_i - \bar{x})^2.\end{aligned}$$

3.4b. For completeness, we include here a review of variance and standard deviation. The *standard deviation* of the random variable $\hat{\beta}$, denoted $\text{SD}(\hat{\beta})$, can be informally defined as follows.

As in Note 3.3d, imagine an infinite number of repetitions of the experiment and imagine calculating an estimate $\hat{\beta}$ from each such experiment. The expectation of the random variable $\hat{\beta}$ is the value around which the imagined collection of $\hat{\beta}$'s would be centered. Since $\hat{\beta}$ is an unbiased estimate, this "central" value is β. But of course the $\hat{\beta}$'s would vary. One way to measure the typical size of the deviations of the $\hat{\beta}$'s from the center is by the *standard deviation*, $\text{SD}(\hat{\beta})$. Perhaps it would seem most straightforward to define $\text{SD}(\hat{\beta})$ to be the average of the infinite collection of absolute deviations $|\hat{\beta} - \beta|$, but for reasons of mathematical convenience, $\text{SD}(\hat{\beta})$ is defined to be the square root of the average of the infinite collection of squared deviations $(\hat{\beta} - \beta)^2$. The average of the squared deviations, without taking a square root, is the *variance*, $\text{Var}(\hat{\beta})$.

3.4c. To say that $\hat{\sigma}^2$ is an *unbiased* estimate of σ^2 means that the expectation of $\hat{\sigma}^2$, viewed as a random variable, is σ^2.

3.4d. For a discussion, of p-values, see Gibbons and Pratt (1975).

3.5a. Some optimal properties of least-squares estimates and tests under the assumption of normally distributed errors are listed in Chapter 9.

3.5b. By saying that a test of $\beta = 0$ is approximately "valid" we mean that the calculated p-value is approximately equal to what it is supposed to be, namely, the probability, assuming $\beta = 0$, that in a repetition of the experiment, the test statistic would be as large or larger than the test statistic observed in the actual experiment. By saying that one test is "more powerful" than another test we mean that, when $\beta \neq 0$, the p-value of the first test tends to be smaller, thus giving a stronger indication that $\beta \neq 0$.

3.5c. For more information on normal probability plots, see Section 3.8 and Appendix 3A in Daniel and Wood (1980), Section 3.1 and Appendix 3A in Draper and Smith (1981), and Section 6.6 in Weisberg (1985). Daniel and Wood show how difficult it is to judge nonnormality from a normal probability plot of the residuals in small samples. For information on tests of normality, see Section 6.6 in Weisberg (1985) and Section 9.6 in D'Agostino and Stephens (1986).

3.7. Matrix notation allows convenient calculation of expectations and variances. Let

$$\mathbf{y} = \begin{bmatrix} y_1 \\ y_2 \\ \vdots \\ y_n \end{bmatrix}$$

be a random vector, that is, a vector whose components are random variables. The *expectation vector* of $\mathbf{y}$ is defined to be

$$E(\mathbf{y}) = \begin{bmatrix} E(y_1) \\ E(y_2) \\ \vdots \\ E(y_n) \end{bmatrix}$$

The *variance–covariance matrix* of $\mathbf{y}$ is defined to be

$$\text{Cov}(\mathbf{y}) = \begin{bmatrix} \text{Var}(y_1) & \text{Cov}(y_1, y_2) & \cdots & \text{Cov}(y_1, y_n) \\ \text{Cov}(y_2, y_1) & \text{Var}(y_2) & \cdots & \text{Cov}(y_2, y_n) \\ \vdots & \vdots & & \vdots \\ \text{Cov}(y_n, y_1) & \text{Cov}(y_n, y_2) & \cdots & \text{Var}(y_n) \end{bmatrix}$$

Suppose A is an $m \times n$ matrix whose entries are constant numbers. Two convenient rules for calculating expectations and variances are

$$\text{(a)}\quad E(A\mathbf{y}) = AE(\mathbf{y})$$

$$\text{(b)}\quad \text{Cov}(A\mathbf{y}) = A\,\text{Cov}(\mathbf{y})A'$$

These are generalizations of the familiar facts that if y is a random variable and a is a constant number, then

$$E(ay) = aE(y)$$
$$\mathrm{Var}(ay) = a^2\,\mathrm{Var}(y)$$

These latter equations are the special cases of (a) and (b) when $m = 1$ and $n = 1$.

To prove these rules in general would involve a lot of subscripts, but we can convince ourselves further by looking at another special case. Let $m = 1$ and $n = 2$. Then

$$\boldsymbol{A} = [a_1 \quad a_2], \qquad \boldsymbol{y} = \begin{bmatrix} y_1 \\ y_2 \end{bmatrix}$$

and

$$\boldsymbol{Ay} = [a_1 \quad a_2]\begin{bmatrix} y_1 \\ y_2 \end{bmatrix} = a_1 y_1 + a_2 y_2$$

So

$$\begin{aligned} E(\boldsymbol{Ay}) &= E(a_1 y_1 + a_2 y_2) = a_1 E(y_1) + a_2 E(y_2) \\ &= [a_1 \quad a_2]\begin{bmatrix} E(y_1) \\ E(y_2) \end{bmatrix} = \boldsymbol{A}E(\boldsymbol{y}) \end{aligned}$$

Also,

$$\begin{aligned} \mathrm{Cov}(\boldsymbol{Ay}) &= \mathrm{Var}(a_1 y_1 + a_2 y_2) \\ &= a_1^2\,\mathrm{Var}(y_1) + 2a_1 a_2\,\mathrm{Cov}(y_1, y_2) + a_2^2\,\mathrm{Var}(y_2) \\ &= [a_1 \quad a_2]\begin{bmatrix} \mathrm{Var}(y_1) & \mathrm{Cov}(y_1, y_2) \\ \mathrm{Cov}(y_2, y_1) & \mathrm{Var}(y_2) \end{bmatrix}\begin{bmatrix} a_1 \\ a_2 \end{bmatrix} \\ &= \boldsymbol{A}\,\mathrm{Cov}(\boldsymbol{y})\boldsymbol{A}' \end{aligned}$$

Let us apply (a) and (b) to the vector of least-squares estimates. Recall that $\hat{\boldsymbol{\beta}} = \boldsymbol{Ay}$, where $\boldsymbol{A} = (\boldsymbol{X}'\boldsymbol{X})^{-1}\boldsymbol{X}'$. Using (a) we calculate

$$\begin{aligned} E(\hat{\boldsymbol{\beta}}) &= E(\boldsymbol{Ay}) = \boldsymbol{A}E(\boldsymbol{y}) \\ &= (\boldsymbol{X}'\boldsymbol{X})^{-1}\boldsymbol{X}'(\boldsymbol{X\beta}) = (\boldsymbol{X}'\boldsymbol{X})^{-1}(\boldsymbol{X}'\boldsymbol{X})\boldsymbol{\beta} \\ &= \boldsymbol{\beta} \end{aligned}$$

This shows that the least-squares estimates are unbiased; that is, $E(\hat{\beta}_j) = \beta_j$ for all j.

Using (b) we calculate

$$\begin{aligned}
\text{Cov}(\hat{\boldsymbol{\beta}}) &= \text{Cov}(\boldsymbol{A}\boldsymbol{y}) = \boldsymbol{A}\,\text{Cov}(\boldsymbol{y})\boldsymbol{A}' \\
&= (\boldsymbol{X}'\boldsymbol{X})^{-1}\boldsymbol{X}'(\sigma^2\boldsymbol{I})\boldsymbol{X}(\boldsymbol{X}'\boldsymbol{X})^{-1} \\
&= \sigma^2(\boldsymbol{X}'\boldsymbol{X})^{-1}(\boldsymbol{X}'\boldsymbol{X})(\boldsymbol{X}'\boldsymbol{X})^{-1} \\
&= \sigma^2(\boldsymbol{X}'\boldsymbol{X})^{-1}
\end{aligned}$$

3.8a. We are using SSR to denote the sum of squares of the residuals. In other books you may find SSR used to denote the sum of squares due to regression, which is $\Sigma(\hat{y}_i - \bar{y})^2$. You may find the sum of squares of the residuals denoted by SSE, standing for the sum of squares due to error.

3.8b. The expected value of the difference between the residual sums of squares is obtained as follows. Note that $\text{SSR}_{\text{full}} = \Sigma\hat{e}_i^2$. Since $\hat{\sigma}^2$ in (3.11) is an unbiased estimate of σ^2, it follows that SSR_{full} is an unbiased estimate of $(n-5)\sigma^2$. Similarly, if the reduced model is true, then $\text{SSR}_{\text{reduced}}$ is an unbiased estimate of $(n-1)\sigma^2$. The 1 in $n-1$ corresponds to the fact that the reduced model has 1 regression coefficient. The expected value of $\text{SSR}_{\text{reduced}} - \text{SSR}_{\text{full}}$ is $(n-1)\sigma^2 - (n-5)\sigma^2 = 4\sigma^2$ when the null hypothesis is true.

3.9. There are two approaches one can take to testing the hypothesis $\beta_{q+1} = \cdots = \beta_p = 0$. The approach we have taken in Sections 3.8 and 3.9 is to compare the sums of squares of the residuals in the full and reduced models. Another approach is to estimate $\beta_{q+1}, \ldots, \beta_p$ and see how close to 0 the estimates are.

To describe the second approach, let $\boldsymbol{\delta} = (\beta_{q+1}, \ldots, \beta_p)$. We want to test whether or not $\boldsymbol{\delta} = \boldsymbol{0}$. The least-squares estimate $\hat{\boldsymbol{\delta}}_{\text{LS}}$ can be obtained as the last $p-q$ entries in $\hat{\boldsymbol{\beta}}_{\text{LS}}$. The variance-covariance matrix of $\hat{\boldsymbol{\delta}}_{\text{LS}}$ is the $(p-q)\times(p-q)$ matrix in the lower right corner of the variance-covariance matrix of $\hat{\boldsymbol{\beta}}_{\text{LS}}$. We know from Note 3.7 that $\text{Cov}(\hat{\boldsymbol{\beta}}_{\text{LS}}) = \sigma^2(\boldsymbol{X}'\boldsymbol{X})^{-1}$. Let V_δ denote $\text{Cov}(\hat{\boldsymbol{\delta}}_{\text{LS}})$; substituting $\hat{\sigma}^2$ from (3.12), we obtain an estimate $\hat{V}_\delta$. A reasonable measure of how close $\boldsymbol{\delta}$ is to $\boldsymbol{0}$ is given by $\hat{\boldsymbol{\delta}}'_{\text{LS}}\hat{V}_\delta^{-1}\hat{\boldsymbol{\delta}}_{\text{LS}}$. The two approaches lead to exactly the same test statistic, because it turns out that test statistic (3.13) can be calculated as $F = \hat{\boldsymbol{\delta}}'_{\text{LS}}\hat{V}_\delta^{-1}\hat{\boldsymbol{\delta}}_{\text{LS}}/(p-q)$.

3.10. To determine $\text{Var}(\hat{\beta}_j)$, use the fact (shown in Note 3.7) that $\text{Cov}(\hat{\boldsymbol{\beta}}) = \sigma^2(\boldsymbol{X}'\boldsymbol{X})^{-1}$. Note that $\text{Var}(\hat{\beta}_j)$ is the $(j+1)$th diagonal entry in $\text{Cov}(\hat{\boldsymbol{\beta}})$.

3.11a. To show that R^2 can be expressed as a function of F, let $S_f = \Sigma(y_i - \hat{y}_i)^2$, the sum of squared residuals in the full model, and let $S_r = \Sigma(y_i - \bar{y})^2$, the sum of squared residuals in the reduced model $Y = \beta_0 + e$ with

$\beta_1 = \beta_2 = \cdots = \beta_p = 0$. Using (3.12) and (3.13) with $q = 0$, we have $F = (S_r - S_f)/(pS_f/(n-p-1)) = ((n-p-1)/p)((S_r/S_f) - 1)$. In the next paragraph we will see that $\Sigma(\hat{y}_i - \bar{y})^2 = S_r - S_f$. Hence $R^2 = (S_r - S_f)/S_r = 1 - (S_f/S_r)$. The expression for F can be rearranged to obtain S_f/S_r as a function of F, which can be substituted into the expression for R^2 to obtain

$$R^2 = 1 - 1/(1 + pF/(n-p-1)).$$

It remains to show $\Sigma(\hat{y}_i - \bar{y})^2 = S_r - S_f$. Writing $y_i - \bar{y} = (y_i - \hat{y}_i) + (\hat{y}_i - \bar{y})$, we see $(y_i - \bar{y})^2 = (y_i - \hat{y}_i)^2 + (\hat{y}_i - \bar{y})^2 + 2(y_i - \hat{y}_i)(\hat{y}_i - \bar{y})$ and hence $\Sigma(y_i - \bar{y})^2 = \Sigma(y_i - \hat{y}_i)^2 + \Sigma(\hat{y}_i - \bar{y})^2 + 2\Sigma(y_i - \hat{y}_i)(\hat{y}_i - \bar{y})$. That is, $S_r = S_f + \Sigma(\hat{y}_i - \bar{y})^2 + 2\Sigma(y_i - \hat{y}_i)(\hat{y}_i - \bar{y})$. It suffices to show $\Sigma(y_i - \hat{y}_i)(\hat{y}_i - \bar{y}) = 0$. Since $\Sigma(y_i - \hat{y}_i)(\hat{y}_i - \bar{y}) = \Sigma(y_i - \hat{y}_i)\hat{y}_i - \bar{y}\Sigma(y_i - \hat{y}_i)$, it suffices to show $\Sigma\hat{y}_i(y_i - \hat{y}_i) = 0$ and $\Sigma(y_i - \hat{y}_i) = 0$. In matrix notation, we want to show $\hat{\mathbf{y}}'(\mathbf{y} - \hat{\mathbf{y}}) = 0$ and $\mathbf{1}'(\mathbf{y} - \hat{\mathbf{y}}) = 0$ where $\mathbf{1}$ is a column of 1's. Recall that $\hat{\mathbf{y}} = \mathbf{X}\hat{\boldsymbol{\beta}}$ and that $\mathbf{1}$ is a column of $\mathbf{X}$. Therefore it suffices to show $\mathbf{X}'(\mathbf{y} - \mathbf{X}\hat{\boldsymbol{\beta}}) = 0$, which follows from (3.8).

3.11b. In simple regression, the coefficient of determination, R^2, can be shown to coincide with the square of the sample correlation between X and Y, r^2. By definition, $r = s_{xy}/(s_x s_y)$, where $s_x^2 = \Sigma(x_i - \bar{x})^2/(n-1)$ is the sample variance of X, s_y^2 is the sample variance of Y, and $s_{xy} = \Sigma(x_i - \bar{x})(y_i - \bar{y})/(n-1)$ is the sample covariance between X and Y. Recall $R^2 = \Sigma(\hat{y}_i - \bar{y})^2/\Sigma(y_i - \bar{y})^2$. In simple regression, $\hat{y}_i - \bar{y} = \hat{\alpha} + \hat{\beta}x_i - \bar{y} = \bar{y} - \hat{\beta}\bar{x} + \beta x_i - \bar{y} = \hat{\beta}(x_i - \bar{x})$. Therefore $R^2 = \hat{\beta}^2\Sigma(x_i - \bar{x})^2/\Sigma(y_i - \bar{y})^2 = \hat{\beta}^2 s_x^2/s_y^2$. From (3.2) we see $\hat{\beta} = s_{xy}/s_x^2$. Hence $R^2 = s_{xy}^2/(s_x^2 s_y^2) = r^2$.

3.12. The formula for $\hat{\beta}$ comes from the algebraic identity $\Sigma(x_i - \bar{x})(y_i - \bar{y}) = \Sigma x_i y_i - (\Sigma x_i)(\Sigma y_i)/n$. This identity can be obtained by noting $\Sigma(x_i - \bar{x})\bar{y} = \bar{y}\Sigma(x_i - \bar{x}) = 0$; hence $\Sigma(x_i - \bar{x})(y_i - \bar{y}) = \Sigma(x_i - \bar{x})y_i = \Sigma x_i y_i - \bar{x}\Sigma y_i = \Sigma x_i y_i - (\Sigma x_i)(\Sigma y_i)/n$. By considering the case $x_i = y_i$, we also obtain the identity $\Sigma(x_i - \bar{x})^2 = \Sigma x_i^2 - (\Sigma x_i)^2/n$.

Additional Reading. To learn more about least-squares regression, see Draper and Smith (1981), Gunst and Mason (1980), Myers (1990), Neter, Wasserman, and Kutner (1989) or Weisberg (1985).

REFERENCES

D'Agostino, R. B., and M. A. Stephens (eds.) (1986). *Goodness-of-Fit Techniques*. Marcel Dekker, New York.

Daniel, C., and F. S. Wood (1980). *Fitting Equations to Data*, 2nd ed. Wiley, New York.

Draper, N. R., and H. Smith (1981). *Applied Regression Analysis*, 2nd ed. Wiley, New York.

Gibbons, J. D., and J. W. Pratt (1975). *P*-values: interpretation and methodology. *American Statistician*, vol. 29, pp. 20–24.

Gunst, R. F., and R. L. Mason (1980). *Regression Analysis and its Application: A Data-Oriented Approach*. Marcel Dekker, New York.

Myers, R. H. (1990). *Classical and Modern Regression with Applications*. PWS-Kent, Boston.

Neter, J., W. Wasserman, and M. H. Kutner (1989). *Applied Linear Regression Models*, 2nd ed. Irwin, Homewood, IL.

Stigler, S. M. (1986). *The History of Statistics: The Measurement of Uncertainty Before 1900*. Harvard University Press, Cambridge, MA.

Weisberg, S. (1985). *Applied Linear Regression*, 2nd ed. Wiley, New York.

CHAPTER 4

Least-Absolute-Deviations Regression

4.1 INTRODUCTION

The method of least absolute deviations was introduced almost 50 years before the method of least squares, in 1757 by Roger Joseph Boscovich. He devised the method as a way to reconcile inconsistent measurements for the purpose of estimating the shape of the earth. After Pierre Simon Laplace adopted the method 30 years later, it saw occasional use, but it was soon overshadowed by the method of least squares.

The popularity of least squares was at least partly due to the relative simplicity of its computations and to the supporting theory that was developed for it by Gauss and Laplace. Today, computation is not such a limitation and theoretical foundations have been laid for a variety of alternative methods, including the method of least absolute deviations (LAD).

4.2 ESTIMATING THE REGRESSION LINE

Consider the data in Table 1.3 consisting of birth rates and urban percentages for 14 North and Central American countries. It was stated in Section 1.6 that the estimate of the regression line obtained by the LAD method is $\hat{Y} = 46 - 0.54X$. Now let us see how this method works.

We begin by proposing the linear regression model $Y = \alpha + \beta X + e$. Then we make a scatter plot as in Figure 1.1. The data point corresponding to Trinidad/Tobago stands apart as a possible outlier, but the main pattern of the data seems to show a trend sloping downward from left to right. So, from this initial appraisal, the linear regression model seems reasonable. Let us estimate the regression coefficients α and β by the method of least absolute deviations.

In the method of least squares, the estimates $\hat{\alpha}$ and $\hat{\beta}$ are chosen so that the sum of the squares of the residuals, $\Sigma \hat{e}_i^2$, is as small as possible. In the method of least absolute deviations, the estimates $\hat{\alpha}$ and $\hat{\beta}$ are chosen so that the sum of the absolute values of the residuals, $\Sigma |\hat{e}_i|$, is as small as possible. That is, the *least-absolute-deviations estimates* $\hat{\alpha}$ and $\hat{\beta}$ are the values of a and b that minimize

$$\sum |y_i - (a + bx_i)| \tag{4.1}$$

The difference $y_i - (a + bx_i)$ is called the *deviation* of the point (x_i, y_i) from the line $\hat{Y} = a + bX$.

The concept of LAD estimation is no more difficult than the concept of least-squares estimation; in fact it is simpler in so far as $|\hat{e}|$ is a more straightforward measure of the size of a residual than is $\hat{e}^2$. But in the actual calculation of the estimates, the LAD method is more complicated. There are no formulas for the LAD estimates; instead we present an algorithm for calculating them. For clarity, the algorithm assumes that the data set is not subject to nonuniqueness or degeneracy. These are two phenomena that can cause technical problems, but they occur only infrequently. They are discussed in the next section.

An Algorithm. Our object is to find the line that best fits the data in the sense of having the least sum of absolute deviations. The main part of the algorithm is a procedure that, for any given point (x_0, y_0), finds the best line among all the lines passing through it. This procedure is used together with the fact that the LAD regression line passes through two of the data points. So the algorithm starts with one of the data points, say (x_1, y_1), and finds the best line passing through it. This line also passes through another data point; by reindexing we can denote this other data point by (x_2, y_2). Next we find the best line passing through (x_2, y_2). This line also passes through another data point; by reindexing we can denote this other data point by (x_3, y_3). Next we find the best line passing through (x_3, y_3), and so on. As the algorithm continues, the lines that are obtained are increasingly better. Eventually, the most recent line obtained will be the same as the previous line. This is the best line among all lines, without regard to what points they pass through. That is, this is the LAD regression line.

Now we need to describe the procedure for finding the best line among all lines passing through a given point (x_0, y_0). For each data point (x_i, y_i) calculate the slope $(y_i - y_0)/(x_i - x_0)$ of the line passing through the two points (x_0, y_0) and (x_i, y_i). If $x_i = x_0$ for some i, the slope is not defined but such points can be ignored. Reindex the data points so that $(y_1 - y_0)/(x_1 - x_0) \leq (y_2 - y_0)/(x_2 - x_0) \leq \cdots \leq (y_n - y_0)/(x_n - x_0)$. Let $T = \Sigma|x_i - x_0|$. Find the index k that satisfies the conditions

$$
\begin{aligned}
|x_1 - x_0| + \cdots + |x_{k-1} - x_0| &< \tfrac{1}{2}T \\
|x_1 - x_0| + \cdots + |x_{k-1} - x_0| + |x_k - x_0| &> \tfrac{1}{2}T
\end{aligned}
\tag{4.2}
$$

The best line passing through (x_0, y_0) is the line $\hat{Y} = \alpha^* + \beta^* X$, where

$$
\begin{aligned}
\beta^* &= \frac{y_k - y_0}{x_k - x_0} \\
\alpha^* &= y_0 - \beta^* x_0
\end{aligned}
\tag{4.3}
$$

Justification That the Line Passes Through Two Data Points. A key fact used in devising the algorithm is that the LAD regression line passes through two of the data points. To see why this is so, imagine a line $y = a + bx$ drawn through a plot of the data points. The absolute deviation of a data point from the line is the length of the vertical line segment from the point to the line. The value of (4.1) is the sum of these absolute deviations. Suppose the line does not pass through any data points. If the line is moved upward a small amount, say ε, then each absolute deviation either decreases by ε or increases by ε, depending on whether the data point is above or below the line. The value of (4.1) can be decreased (or at least not increased) by moving the line either upward or downward, depending on whether there are more data points above or below the line. Move the line until it meets a data point.

If the line passes through exactly one data point, the line can be pivoted on its data point, either clockwise or counterclockwise, until it meets a second data point. Of course the absolute deviation for the one data point remains zero, and each of the other absolute deviations either decreases or increases (by varying amounts). With respect to decreasing or increasing the sum of absolute deviations, pivoting clockwise and pivoting counterclockwise have opposite effects. So by pivoting one way or the other, we can decrease (or at least not increase) the sum of absolute deviations. This shows that, in order to minimize the sum of absolute deviations, we need only look at lines that pass through at least two of the data points.

Justification of the Algorithm. The procedure for finding the best line passing through (x_0, y_0) can be justified as follows. Among all lines passing through (x_0, y_0) we want to find a line that minimizes (4.1). For a line $\hat{Y} = a + bX$ to pass through (x_0, y_0) means that $y_0 = a + bx_0$, hence $a = y_0 - bx_0$. This justifies the second part of (4.3).

The deviation $y_i - (a + bx_i)$ can now be written as $(y_i - y_0) - b(x_i - x_0)$. So we want to find the value of b that minimizes

$$\sum |(y_i - y_0) - b(x_i - x_0)| \tag{4.4}$$

Consider (4.4) as a function of b. A common technique for finding the minimum of a function is to differentiate the function. Although the absolute value function $|t|$ is not differentiable at $t = 0$, it is differentiable at all $t \neq 0$. This allows us to differentiate (4.4) at all b except for b satisfying $(y_i - y_0) - b(x_i - x_0) = 0$, that is, except for the n values $b_i = (y_i - y_0)/(x_i - x_0)$. The inequalities (4.2) express the condition that the derivative of (4.4) is negative for $b < \beta^*$ and is positive for $b > \beta^*$. Together with the fact that (4.4) is a continuous function, even at the points where it is not differentiable, this implies that (4.4) is decreasing for $b < \beta^*$ and is increasing for $b > \beta^*$, which implies that β^* is the minimizing value of b. This justifies the first part of (4.3).

Note that the best line passing through (x_0, y_0) passes through one of the data points, namely, (x_k, y_k). To check this, one can use the definitions of α^* and β^* in (4.3) to show that $y_k = \alpha^* + \beta^* x_k$. This confirms the statement that the best line passing through one of the data points also passes through another data point, which was justified in the preceding subsection.

Recall that the slope $(y_i - y_0)/(x_i - x_0)$ is not defined when $x_i = x_0$. It is permissible to ignore such data points because, if the line is required to pass through the given point (x_0, y_0), then a data point with the same x-value as this point can have no influence in determining the best line. This is because its contribution $|(y_i - y_0) - b(x_i - x_0)|$ to the sum (4.4) has the same value $|y_i - y_0|$ for all b.

The Birth Rate Data. Let us apply the LAD simple regression algorithm to the birth rate data in Table 1.3. As a first step we find the best line passing through the Canada data point (55.0, 16.2). To do this, we form the slopes $(y_i - 16.2)/(x_i - 55.0)$. For example, the slope for Costa Rica is $(30.5 - 16.2)/(27.3 - 55.0) = -0.5162$. The slopes for the 13 countries other than Canada are shown in Table 4.1. They are listed in increasing order.

To apply (4.2), calculate $\sum|x_i - 55.0| = 355.9$, divide it by 2 to obtain $355.9/2 = 177.95$, and look for the country for which the cumulative sum in column 4 of Table 4.1 first exceeds 177.95. We see that this is El Salvador, because $172.5 < 177.95$ and $216.0 > 177.95$. Therefore $\beta^* = -0.5517$ and $\alpha^* = 16.2 - (-0.5517)(55.0) = 46.54$. Actually, we do not need to calculate α^* and β^* at this step; we only need to know that the best line through the Canada data point also passes through the El Salvador data point.

Table 4.1 Calculations Used in Finding the Best Line Through the Canada Data Point

Country	$\frac{y_i - 16.2}{x_i - 55.0}$	$\|x_i - 55.0\|$	Cumulative Sum of $\|x_i - 55.0\|$
Mexico	−1.5000	11.8	11.8
Nicaragua	−1.0566	26.5	38.3
Dominican Republic	−0.9441	17.9	56.2
Honduras	−0.7694	36.0	92.2
Panama	−0.6821	17.3	109.5
Haiti	−0.6107	41.1	150.6
Jamaica	−0.5525	21.9	172.5
El Salvador	−0.5517	43.5	216.0
Guatemala	−0.5447	40.8	256.8
Costa Rica	−0.5162	27.7	284.5
Trinidad/Tobago	−0.1743	48.2	332.7
United States	−0.1333	1.5	334.2
Cuba	−0.0323	21.7	355.9

The next step is to find the best line passing through the El Salvador data point. We form the slopes $(y_i - 40.2)/(x_i - 11.5)$, put them in increasing order, and construct a table similar to Table 4.1 with a column for $|x_i - 11.5|$ and a column for the cumulative sums. If you construct this table, you will find that the total cumulative sum is 265.5 and that the country whose cumulative sum first exceeds $265.5/2 = 132.75$ is the United States. Therefore the best line passing through the El Salvador data point also passes through the United States data point.

The next step is to find the best line passing through the United States data point. Constructing a table similar to Table 4.1 and looking at the cumulative sums, we find that the best line passing through the United States data point also passes through the El Salvador data point. But this is the same line we obtained at the previous step, and so the algorithm stops. The LAD regression line is the line passing through the data points of El Salvador and the United States. Its slope is $\hat{\beta} = (40.2 - 16.0)/(11.5 - 56.5) = -0.5378$ and its intercept is $\hat{\alpha} = 40.2 - (-0.5378)(11.5) = 46.38$. That is, the LAD regression line is $\hat{Y} = 46.38 - 0.5378X$. See Figure 1.1.

4.3 NONUNIQUENESS AND DEGENERACY

The algorithm described in the preceding section will find the LAD regression line for most data sets, but occasionally problems arise with nonuniqueness or degeneracy. Nonuniqueness means that there is more than one best line passing through a data point. Degeneracy means that the best line through a data point also passes through two or more other data points. Recall that the algorithm proceeds in steps. At each step we find the best line passing through a given data point. The best line always passes through another data point, and this data point is used in the next step. But when there is nonuniqueness, there is more than one best line. And when there is degeneracy, the best line passes through more than one other data point. In either case, there is more than one choice for the data point to be used in the next step. By making unlucky choices, the algorithm may go around in circles, or it may stop with a line that is not the LAD regression line.

The possibility of such problems is indicated either when equality occurs in condition (4.2) or when the slope $\beta^* = (y_k - y_0)/(x_k - x_0)$ in (4.3) is equal to $(y_{k-1} - y_0)/(x_{k-1} - x_0)$ or $(y_{k+1} - y_0)/(x_{k+1} - x_0)$. In such instances we can resort to the algorithm in the following paragraph. Also see the subsection on nonuniqueness and degeneracy at the end of Section 4.6.

A Simpler Algorithm. Another possible approach is to replace the algorithm in Section 4.2 by the following algorithm, which has the advantage of being conceptually simple and the disadvantage of requiring more computation. It is known that the LAD regression line (or at least one of them, in case of nonuniqueness) passes through at least two data points. So an LAD regression line can be found among the lines defined by all possible pairs of data points. (Some of these lines may coincide.) We can simply compute the sum of absolute deviations (4.1) for each of these lines and choose the one (or ones) with the smallest sum. The feasibility of this algorithm depends on the sample size n.

This algorithm is not disturbed by degeneracy. In case of nonuniqueness, when there are several LAD regression lines, we could arbitrarily choose one, or we could take their average. This average line is also an LAD regression line.

4.4 TESTING $\beta = 0$

The method of least absolute deviations estimates the slope of the regression line for the birth rate data to be $\hat{\beta} = -0.5378$, but we do not expect the estimate to be exactly equal to the true value. Even though $\hat{\beta}$ is negative, it is possible that the true value of β may actually be 0. That is, what appears to be a negative relationship between birth rate and urban percentage may be due merely to the randomness of the data. Let us test whether or not β could be 0.

Table 4.2 The Nonzero Residuals from the LAD Regression Analysis of the Birth Rate Data, Arranged in Increasing Order

i	$\hat{e}_{(i)}$
1	−18.128
2	−11.576
3	−1.203
4	−0.607
5	−0.348
6	−0.284
7	1.890
8	2.391
9	6.667
10	7.733
11	10.748
12	13.142

Description of the Test. First calculate the LAD regression estimates $\hat{\alpha}$ and $\hat{\beta}$ and the residuals $\hat{e}_i = y_i - (\hat{\alpha} + \hat{\beta}x_i)$. Let $m = n - 2$, the number of nonzero residuals. Arrange the nonzero residuals in increasing order. Let $\hat{e}_{(1)}$ denote the smallest residual, $\hat{e}_{(2)}$ the next smallest, ..., and $\hat{e}_{(m)}$ the largest. Table 4.2 displays them for the birth rate data.

Let k_1 be the closest integer to $(m + 1)/2 - \sqrt{m}$ and let k_2 be the closest integer to $(m + 1)/2 + \sqrt{m}$. Calculate

$$\hat{\tau} = \frac{\sqrt{m}\left[\hat{e}_{(k_2)} - \hat{e}_{(k_1)}\right]}{4} \tag{4.5}$$

Then calculate

$$\text{est.SD}(\hat{\beta}) = \frac{\hat{\tau}}{\sqrt{\sum(x_i - \bar{x})^2}} \tag{4.6}$$

The test statistic is

$$|t| = \frac{|\hat{\beta}|}{\text{est.SD}(\hat{\beta})} \tag{4.7}$$

The p-value of the test is calculated as the probability $\text{Prob}[|T| \geq |t|]$ where T denotes a random variable having a t distribution with $n - 2$ degrees of freedom.

Comparison with the Least-Squares Test. The LAD test described above is similar to the least-squares test. In fact, formula (4.7) is exactly the same as formula (3.5). But $\hat{\beta}$ and est.SD($\hat{\beta}$) are calculated differently. No exact formula for SD($\hat{\beta}$) is available in LAD regression. The only difference in the two formulas for est.SD($\hat{\beta}$) is that $\hat{\sigma}$ appears in the least-squares formula and $\hat{\tau}$ appears in the LAD formula.

Justification of (4.6) and (4.7). Test statistic (4.7) is sensible to use for testing whether $\beta = 0$. The estimate $\hat{\beta}$ can be expected to be fairly close to β; the distance between $\hat{\beta}$ and β should not be more than one or two standard deviations SD($\hat{\beta}$). If $|t|$ is large, this means that the distance between $\hat{\beta}$ and 0 is much larger than est.SD($\hat{\beta}$), and so we conclude $\beta \neq 0$.

Formula (4.6) makes sense in so far as it implies that SD($\hat{\beta}$) is estimated to be smaller when the x-values are more spread out. A wider range of x-values gives better information about the slope of the regression line. Also, estimation is more precise when τ is small. It remains to discuss τ.

The Parameter τ. The quantity $\hat{\tau}$ in (4.5) is an estimate of a parameter τ which plays a role in LAD regression analogous to the role of σ in least-squares regression. The standard deviation of $\hat{\beta}_{\text{LS}}$ is $\sigma/\sqrt{\Sigma(x_i - \bar{x})^2}$, whereas the standard deviation of $\hat{\beta}_{\text{LAD}}$ is approximately $\tau/\sqrt{\Sigma(x_i - \bar{x})^2}$ (the approximation is better for large sample sizes). Thus the ratio τ/σ determines which of the two regression methods is better for estimating the slope of the regression line.

Both τ and σ are measures of the size of the random errors. τ is equal to $1/(2\theta)$, where θ is the probability density of the error distribution at its median. (When doing LAD regression, the assumption is sometimes made that the median, rather than the mean, of the error distribution is 0. If the error distribution is symmetric, its mean and median coincide.) If σ is large, then the errors are widely spread out, so the probability density near the median is low, hence θ is small, which implies τ is large.

So, roughly speaking, τ is large when σ is large and τ is small when σ is small. But the exact ratio τ/σ depends on the shape of the distribution of the population of errors. If the errors have a normal distribution, then $\tau/\sigma = 1.253 > 1$, and so, at least for large sample sizes, the LAD regression estimates are less accurate then the least-squares regression estimates. If the errors have a Laplace distribution, then $\tau/\sigma = 0.707 < 1$.

Justification of (4.5). Our justification assumes a large sample size n. Then m is approximately equal to n and the residuals $\hat{e}_i$ behave much the same as the true errors e_i. We will argue that τ can be approximated by

$\sqrt{n}\,[\hat{e}_{(k_2)} - \hat{e}_{(k_1)}]/4$ where k_1 and k_2 are the closest integers to $(n + 1)/2 \pm \sqrt{n}$. Our argument consists of deriving two approximate 95%-confidence intervals for the median of the error distribution, both based on normal approximations, and equating their lengths.

Recall from the preceding subsection that τ in LAD regression is analogous to σ in least-squares regression. The standard deviation of an LS regression estimate has the form σc, whereas the standard deviation of the corresponding LAD regression estimate is approximately τc for the same quantity c. In particular, consider the special case of "regression" when $p = 0$, that is, when there are no explanatory variables and we simply have a random sample of n independent observations from the same population, such as $e_1, e_2, \ldots, e_n$. The LS "regression" estimate is the sample mean $\bar{e}$, because $a = \bar{e}$ is the value of a that gives the least sum of squared deviations $\Sigma(e_i - a)^2$. Its standard deviation is $\sigma/\sqrt{n}$. The LAD "regression" estimate is the sample median $\tilde{e}$, because $a = \tilde{e}$ is the value of a that gives the least sum of absolute deviations $\Sigma|e_i - a|$. Its standard deviation is approximately $\tau/\sqrt{n}$.

The central limit theorem says that, for large sample sizes n, the distribution of the sample mean is approximately normal. It turns out that for large n the distribution of the sample median is also approximately normal. Also, the expectation of the sample median is approximately equal to the population median. Letting ν denote the median of the population of errors, we can state that, for large n, $\tilde{e}$ has approximately a normal distribution with mean ν and standard deviation $\tau/\sqrt{n}$. Hence, if $\hat{\tau}$ is an estimate of τ, we can construct an approximate 95%-confidence interval for ν to be $\tilde{e} \pm 2\hat{\tau}/\sqrt{n}$. We are not actually interested in ν and are not interested in the confidence interval itself but only in its length, $4\hat{\tau}/\sqrt{n}$.

Next, it can be shown that the interval from $e_{(k_1)}$ to $e_{(k_2)}$ is also an approximate 95%-confidence interval for ν. It turns out that the two intervals are similar, at least for large sample sizes. In particular, their lengths are similar: $4\hat{\tau}/\sqrt{n} \approx e_{(k_2)} - e_{(k_1)}$. Choosing $\hat{\tau}$ to make this an exact equality, replacing e by $\hat{e}$, and replacing n by $m = n - 2$, we obtain formula (4.5). It has been found that using only the m nonzero residuals improves the performance of the test for small samples.

The Distribution of t. The LAD test procedure assumes that, when $\beta = 0$, the random variable t has a t distribution with $n - 2$ degrees of freedom. This is not exactly true but theorems have been proved which imply that when n is large and $\beta = 0$, then t has approximately a standard normal distribution. This justifies the LAD test procedure when n is large, because a t distribution with many degrees of freedom is very similar to the standard normal distribution. When n is small, to be on the safe side so that we do not reject the null hypothesis too hastily, we use the t distribution rather than the standard normal distribution.

The Birth Rate Data. With a sample size as small as $n = 14$, we should be cautious about the validity of the LAD test described in this section, but as an illustration, let us apply the test to the birth rate data.

In Table 4.2 the ordered nonzero residuals are listed. Since $m = n - 2 = 12$ and $(m + 1)/2 - \sqrt{m} = 13/2 - \sqrt{12} = 3.04$, we have $k_1 = 3$. Similarly, $k_2 = 10$. So $\hat{\tau} = \sqrt{12}[\hat{e}_{(10)} - \hat{e}_{(3)}]/4$. In Table 4.2 we see $\hat{e}_{(3)} = -1.203$ and $\hat{e}_{(10)} = 7.733$. Hence $\hat{\tau} = 7.739$. Next we calculate $\Sigma(x_i - \bar{x})^2 = 3151$. So est.SD($\hat{\beta}$) $= 7.739/\sqrt{3151} = 0.1379$, and $|t| = |-0.5378|/0.1379 = 3.900$. To calculate the p-value we use the t distribution with $n - 2 = 12$ degrees of freedom. From the t table in the Appendix, we find that the p-value is between 0.001 and 0.01. We conclude that the slope of the regression line is truly negative.

As a check on the LAD test, let us perform the LS test. In Figure 1.1 note the position of the Trinidad/Tobago data point. We saw in Section 1.6 that this point has a disproportionate influence on the slope of the LS regression line. Note that its influence is in the direction of pulling the slope toward 0. So in the presence of such influence, if the LS test concludes $\beta \neq 0$, we can believe it. When the LS test is applied to these data, it yields a p-value of 0.018. This still indicates $\beta \neq 0$, but not as strongly as the LAD test does.

4.5 AN EXAMPLE OF MULTIPLE REGRESSION

Let us apply the method of least absolute deviations to analyze the data in Table 4.3 concerning the incidence of fires in a residential area. We want to see how the incidence of fires is related to three characteristics of the area: the age of its houses, its incidence of theft, and the income of its families. The data are for 47 predominantly residential areas in Chicago for the year 1975. The column labeled FIRE lists the number of fires per 1000 housing units in the area; the column labeled AGE lists the proportion of housing units built before 1940; the column labeled THEFT lists the number of thefts per 1000 residents; and the column labeled INCOME lists the median family income as a multiple of \$1000.

These data are analyzed in Example 2 in Section 2.4. In the first part of this analysis, it is found that areas 7 and 24 are leverage points and should be deleted from the model. Next, a problem with unequal variances is corrected by transforming FIRE to log(FIRE). At this point in the analysis the model

Table 4.3 Fire Data

Area	FIRE	log(FIRE)	AGE	THEFT	INCOME
1	6.2	1.825	0.604	29	11.744
2	9.5	2.251	0.765	44	9.323
3	10.5	2.351	0.735	36	9.948
4	7.7	2.041	0.669	37	10.656
5	8.6	2.152	0.814	53	9.730
6	34.1	3.529	0.526	68	8.231
7	11.0	2.398	0.426	75	21.480
8	6.9	1.932	0.785	18	11.104
9	7.3	1.988	0.901	31	10.694
10	15.1	2.715	0.898	25	9.631
11	29.1	3.371	0.827	34	7.995
12	2.2	0.788	0.402	14	13.722
13	5.7	1.740	0.279	11	16.250
14	2.0	0.693	0.077	11	13.686
15	2.5	0.916	0.638	22	12.405
16	3.0	1.099	0.512	17	12.198
17	5.4	1.686	0.851	27	11.600
18	2.2	0.788	0.444	9	12.765
19	7.2	1.974	0.842	29	11.084
20	15.1	2.715	0.898	30	10.510
21	16.5	2.803	0.727	40	9.784
22	18.4	2.912	0.729	32	7.342
23	36.2	3.589	0.631	41	6.565
24	39.7	3.681	0.830	147	7.459
25	18.5	2.918	0.783	22	8.014
26	23.3	3.148	0.790	29	8.177
27	12.2	2.501	0.480	46	8.212
28	5.6	1.723	0.715	23	11.230
29	21.8	3.082	0.731	4	8.330
30	21.6	3.073	0.650	31	5.583
31	9.0	2.197	0.754	39	8.564
32	3.6	1.281	0.208	15	12.102
33	5.0	1.609	0.618	32	11.876
34	28.6	3.353	0.781	27	9.742
35	17.4	2.856	0.686	32	7.520
36	11.3	2.425	0.734	34	7.388
37	3.4	1.224	0.020	17	13.842
38	11.9	2.477	0.570	46	11.040
39	10.5	2.351	0.559	42	10.332
40	10.7	2.370	0.675	43	10.908
41	10.8	2.380	0.580	34	11.156
42	4.8	1.569	0.152	19	13.323
43	10.4	2.342	0.408	25	12.960
44	15.6	2.747	0.578	28	11.260
45	7.0	1.946	0.114	3	10.080
46	7.1	1.960	0.492	23	11.428
47	4.9	1.589	0.466	27	13.731

Source: Andrews and Herzberg (1985, p. 409).

being considered is

$$Y = \beta_0 + \beta_1 X_1 + \beta_2 X_2 + \beta_3 X_3 + e$$

where $Y = \log(\text{FIRE})$, $X_1 = \text{AGE}$, $X_2 = \text{THEFT}$, and $X_3 = \text{INCOME}$. Let us use the method of least absolute deviations to estimate and test the coefficients in this model.

4.6 ESTIMATING THE REGRESSION COEFFICIENTS

The *LAD estimates* $\hat{\beta}_0, \hat{\beta}_1, \hat{\beta}_2, \hat{\beta}_3$ are chosen so that the sum of the absolute values of the residuals, $\sum|\hat{e}_i|$, is as small as possible. That is, $\hat{\beta}_0, \hat{\beta}_1, \hat{\beta}_2, \hat{\beta}_3$ are the values of b_0, b_1, b_2, b_3 that minimize

$$\sum |y_i - (b_0 + b_1 x_{i1} + b_2 x_{i2} + b_3 x_{i3})| \tag{4.8}$$

There are no formulas for the minimizing values, but we will describe an algorithm that can be used to obtain them. The algorithm assumes that the data set does not involve any problem with nonuniqueness or degeneracy. These problems are discussed at the end of the section.

Outline of the Algorithm. The presentation of the algorithm is facilitated by using vector notation. Let

$$\boldsymbol{b} = \begin{bmatrix} b_0 \\ b_1 \\ b_2 \\ b_3 \end{bmatrix} \quad \text{and} \quad \boldsymbol{x}_i = \begin{bmatrix} 1 \\ x_{i1} \\ x_{i2} \\ x_{i3} \end{bmatrix}$$

Then the sum of absolute deviations in (4.8) can be written as

$$\sum |y_i - \boldsymbol{b}'\boldsymbol{x}_i| \tag{4.9}$$

We want to find the vector $\boldsymbol{b}$ that minimizes (4.9).

Recall that the algorithm for simple LAD regression is iterative. That is, we start with one line, then we find a better line, then an even better line, and so on until we obtain the best line. Similarly, the algorithm for multiple LAD regression is iterative. We start with a vector $\boldsymbol{b}$, then we find a better vector (in the sense of giving a smaller value for (4.9)), and so on until we obtain the best vector $\hat{\boldsymbol{\beta}}$. At each step, having a vector of estimates $\boldsymbol{b}$, we find a better vector $\boldsymbol{b}^*$ by first finding a suitable "direction" vector $\boldsymbol{d}$ and then finding the value of t for which $\boldsymbol{b}^* = \boldsymbol{b} + t\boldsymbol{d}$ is best.

Finding the Best Vector of Estimates in Direction $\boldsymbol{d}$. We need a procedure for finding the value of t that minimizes

$$\sum |y_i - (\boldsymbol{b} + t\boldsymbol{d})'\boldsymbol{x}_i| \tag{4.10}$$

If we write $z_i = y_i - \boldsymbol{b}'\boldsymbol{x}_i$ and $w_i = \boldsymbol{d}'\boldsymbol{x}_i$, then the procedure must find the value of t that minimizes

$$\sum |z_i - tw_i| \tag{4.11}$$

This is the same as the problem of finding b that minimizes (4.4), for which we know the solution. We take the ratios z_i/w_i, put them in increasing order, reindex the z's and w's according to this order, and find the index k that satisfies

$$|w_1| + |w_2| + \cdots + |w_{k-1}| < \tfrac{1}{2}T$$

$$|w_1| + |w_2| + \cdots + |w_{k-1}| + |w_k| > \tfrac{1}{2}T$$

where $T = \Sigma|w_i|$. The minimizing value of t is z_k/w_k.

Finding a Suitable Direction. At each step the algorithm considers four direction vectors $\boldsymbol{d}_1, \boldsymbol{d}_2, \boldsymbol{d}_3, \boldsymbol{d}_4$. (In the general case with p explanatory variables, it considers $p + 1$ vectors.) These represent eight different directions since for each vector $\boldsymbol{d}_j$ we look in the negative direction $-\boldsymbol{d}_j$ as well as the positive direction $\boldsymbol{d}_j$. Among these eight directions, the most promising direction is the one for which the value of (4.10) decreases most steeply near $t = 0$. To determine how steeply (4.10) decreases, we calculate its right-hand derivative at $t = 0$. In terms of the notation in (4.11), the right-hand derivative at $t = 0$ is $W_- + W_0 - W_+$, where W_- is the sum of $|w_i|$ for indices i for which z_i/w_i is negative, W_0 is the sum of $|w_i|$ for which $z_i = 0$, and W_+ is the sum of $|w_i|$ for which z_i/w_i is positive. We calculate this derivative for each of the eight directions and choose the most suitable direction to be the one whose derivative is most negative. If all the derivatives are positive, then the current vector $\boldsymbol{b}$ is the best vector $\hat{\boldsymbol{\beta}}$, and the algorithm stops.

Starting the Algorithm. Recall that in simple regression, the LAD regression line passes through two of the data points. Similarly, in multiple regression with p explanatory variables, the LAD regression equation is satisfied for $p + 1$ of the data points. In our example, $p = 3$, so $p + 1 = 4$. So a reasonable way to begin the algorithm is to choose four data points, for example the points indexed by $i = 1, 2, 3, 4$, and let the vector of estimates $\boldsymbol{b}$ at the initial step be determined by $y_i = \boldsymbol{b}'\boldsymbol{x}_i$ for $i = 1, 2, 3, 4$. In matrix notation these four equations can be written as $\boldsymbol{Ab} = \boldsymbol{c}$, where

$$\boldsymbol{A} = \begin{bmatrix} \boldsymbol{x}'_1 \\ \boldsymbol{x}'_2 \\ \boldsymbol{x}'_3 \\ \boldsymbol{x}'_4 \end{bmatrix} \quad \text{and} \quad \boldsymbol{c} = \begin{bmatrix} y_1 \\ y_2 \\ y_3 \\ y_4 \end{bmatrix}$$

Therefore, $\boldsymbol{b} = \boldsymbol{A}^{-1}\boldsymbol{c}$. Let the initial set of direction vectors $\boldsymbol{d}_1, \boldsymbol{d}_2, \boldsymbol{d}_3, \boldsymbol{d}_4$ be the four columns of $\boldsymbol{A}^{-1}$.

Iteration of the Algorithm. At each step we have a current vector of estimates $\boldsymbol{b}$ such that the estimated regression equation is satisfied by four data points, say, the points indexed by i_1, i_2, i_3, i_4. Let

$$\boldsymbol{A} = \begin{bmatrix} \boldsymbol{x}'_{i_1} \\ \boldsymbol{x}'_{i_2} \\ \boldsymbol{x}'_{i_3} \\ \boldsymbol{x}'_{i_4} \end{bmatrix}$$

The current set of direction vectors are the four columns of $\boldsymbol{A}^{-1}$.

As described above, we calculate eight derivatives in the directions represented by the four direction vectors and their negatives, and we choose the direction with the most negative derivative. Suppose this is direction $\boldsymbol{d}_3$. A vector of estimates that is better than $\boldsymbol{b}$ is obtained as $\boldsymbol{b}^* = \boldsymbol{b} + t^*\boldsymbol{d}_3$, where t^* is the value of t that minimizes (4.10). In the description of how the minimizing value of t is obtained, note that $t^* = (y_k - \boldsymbol{b}'\boldsymbol{x}_k)/\boldsymbol{d}'_3\boldsymbol{x}_k$ for some k. Replace the third row (because the third direction vector is being used) of $\boldsymbol{A}$ by $\boldsymbol{x}'_k$. Call this new matrix $\boldsymbol{A}^*$. The new set of direction vectors are the four columns of $\boldsymbol{A}^{*-1}$.

Iteration is continued until a step is reached at which all the derivatives in the eight directions are positive.

Justification That the LAD Equation is Satisfied by $p + 1$ Data Points. A key fact underlying the algorithm is that the LAD equation is satisfied by $p + 1$ data points. The equation $\hat{Y} = a + bX$ is often identified with the line consisting of all points $(X, \hat{Y})$ that satisfy it. Similarly, the equation $\hat{Y} = b_0 + b_1X_1 + \cdots + b_pX_p$ can be identified with the set of all $(p + 1)$-dimensional points $(X_1, \ldots, X_p, \hat{Y},)$ (not just data points) that satisfy it. Such a set is called a "hyperplane". When $p = 1$, the hyperplane is a line in 2-dimensional space, and when $p = 2$, the hyperplane is an plane in 3-dimensional space. In the same way that a line is completely determined by specifying two of its points and a plane is completely determined by specifying three of its points, a hyperplane is completely determined by specifying $p + 1$ of its points. (This geometric fact corresponds to the algebraic fact that the $p + 1$ coefficients $b_0, b_1, \ldots, b_p$ are determined by $p + 1$ equations.)

If a hyperplane contains fewer than $p + 1$ data points, it can be "pivoted" about these points until it meets another data point. By pivoting in an appropriate direction, the sum of absolute deviations of the data points from the hyperplane can be decreased (or at least not increased). (This is easiest to picture in the case $p = 1$, which is discussed in Section 4.2) Therefore, in order to minimize the sum of absolute deviations, we need only look at hyperplanes that pass through at least $p + 1$ of the data points.

Justification of the Algorithm. In light of the preceding subsection, a reasonable algorithm would be to look at a sequence of hyperplanes, each hyperplane being determined by a set of $p + 1$ data points. Given the current hyperplane and its set of determining data points, we could select one data point to remove from the set and select another data point to replace it. The selection should be done so that the hyperplane determined by the new set of data points has a smaller sum of absolute deviations. Actually, the algorithm we presented can be described in this way. Finding a suitable direction amounts to selecting a data point to remove, and finding the best value of t in (4.10) amounts to selecting a replacement.

To see this, consider the case $p = 3$ and suppose the current set of data points is $\boldsymbol{x}_1, \boldsymbol{x}_2, \boldsymbol{x}_3, \boldsymbol{x}_4$. Let

$$\boldsymbol{A} = \begin{bmatrix} \boldsymbol{x}_1' \\ \boldsymbol{x}_2' \\ \boldsymbol{x}_3' \\ \boldsymbol{x}_4' \end{bmatrix} \quad \text{and} \quad \boldsymbol{c} = \begin{bmatrix} y_1 \\ y_2 \\ y_3 \\ y_4 \end{bmatrix}$$

so that the current coefficient vector is $\boldsymbol{b} = \boldsymbol{A}^{-1}\boldsymbol{c}$. If data point $\boldsymbol{x}_3$ is removed and replaced by $\boldsymbol{x}_k$, the new coefficient vector is $\boldsymbol{b}^* = \boldsymbol{A}^{*-1}\boldsymbol{c}^*$ where

$$\boldsymbol{A}^* = \begin{bmatrix} \boldsymbol{x}_1' \\ \boldsymbol{x}_2' \\ \boldsymbol{x}_k' \\ \boldsymbol{x}_4' \end{bmatrix} \quad \text{and} \quad \boldsymbol{c}^* = \begin{bmatrix} y_1 \\ y_2 \\ y_k \\ y_4 \end{bmatrix}$$

This amounts to using direction vector $\boldsymbol{d}_3$ and finding $t^* = (y_k - \boldsymbol{b}'\boldsymbol{x}_k)/\boldsymbol{d}_3'\boldsymbol{x}_k$, because $\boldsymbol{b} + t^*\boldsymbol{d}_3 = \boldsymbol{A}^{*-1}\boldsymbol{c}^*$, which can be verified by using the fact that $\boldsymbol{d}_3$ is the third column of $\boldsymbol{A}^{-1}$. This justifies the use of the columns of $\boldsymbol{A}^{-1}$ as direction vectors.

Of course it is sensible to choose the direction vector that gives the most negative derivative for (4.10) since this promises the most decrease in the sum of absolute deviations. The procedure for finding t^* is justified in Section 4.2.

The Fire Data. To apply the LAD multiple regression algorithm to the fire data in Table 4.3, we start by choosing four of the areas, say, areas 1, 2, 3, and 4. Using the data for these four areas, form

$$\boldsymbol{A} = \begin{bmatrix} 1 & 0.604 & 29 & 11.744 \\ 1 & 0.765 & 44 & 9.323 \\ 1 & 0.735 & 36 & 9.948 \\ 1 & 0.669 & 37 & 10.656 \end{bmatrix} \quad \text{and} \quad \boldsymbol{c} = \begin{bmatrix} 1.825 \\ 2.251 \\ 2.351 \\ 2.041 \end{bmatrix}$$

The initial vector of estimates is

$$\boldsymbol{b} = \boldsymbol{A}^{-1}\boldsymbol{c} = \begin{bmatrix} 47.93 \\ -23.26 \\ -0.1161 \\ -2.443 \end{bmatrix}$$

and the initial matrix of direction vectors is

$$\boldsymbol{A}^{-1} = \begin{bmatrix} -284.9 & -308.3 & 157.8 & 436.4 \\ 164.5 & 176.6 & -79.71 & -261.3 \\ 0.5233 & 0.6744 & -0.4656 & -0.7321 \\ 14.59 & 15.51 & -8.185 & -21.91 \end{bmatrix}.$$

Now we want to find a vector of estimates that is better than $\boldsymbol{b}$. To do this we calculate the right-hand derivative of (4.10) for the eight directions represented by the four columns of $\boldsymbol{A}^{-1}$ and their negatives. Denote the first column of $\boldsymbol{A}^{-1}$ by $\boldsymbol{d}_1$. Calculations for obtaining the derivative in the direction $\boldsymbol{d}_1$ are shown in Table 4.4. Recall that areas 7 and 24 have been omitted from the model. Note that $z_i = y_i - \boldsymbol{b}'\boldsymbol{x}_i$ and $w_i = \boldsymbol{d}_1'\boldsymbol{x}_i$. For example,

$$z_5 = 2.152 - [47.93 \quad -23.26 \quad -0.1161 \quad -2.443] \begin{bmatrix} 1 \\ 0.814 \\ 53 \\ 9.730 \end{bmatrix}$$

$$= 3.08$$

Table 4.4 Calculations for Obtaining the Right-Hand Derivative of (4.10) at $t = 0$

i	z_i	w_i	$\text{sign}(z_i/w_i)$	$\lvert w_i \rvert$
1	0	1	0	1
2	0	0	*	0
3	0	0	*	0
4	0	0	*	0
5	3.08	18.71	+	18.71
6	−4.16	−42.68	+	42.68
8	1.48	15.67	+	15.67
9	4.74	35.57	+	35.57
10	2.10	16.43	+	16.43
11	−1.84	−14.42	+	14.42
12	−2.65	−11.21	+	11.21
13	1.27	3.88	+	3.88
14	−10.74	−66.75	+	66.75
15	0.68	12.57	+	12.57
16	−3.15	−13.78	+	13.78
17	5.02	38.48	+	38.48
18	−4.59	−20.88	+	20.88
19	4.07	30.51	+	30.51
20	4.83	31.87	+	31.87
21	0.33	−1.62	−	1.62
22	−6.41	−41.11	+	41.11
23	−8.87	−63.85	+	63.85
25	−4.67	−27.65	+	27.65
26	−3.06	−20.46	+	20.46
27	−8.86	−62.04	+	62.04
28	0.53	8.62	+	8.62
29	−7.03	−41.01	+	41.01
30	−12.50	−80.29	+	80.29
31	−2.74	−15.50	+	15.50
32	−10.51	−66.23	+	66.23
33	0.78	6.80	+	6.80
34	0.52	−0.15	−	0.15
35	−7.03	−45.58	+	45.58
36	−6.44	−38.57	+	38.57
37	−10.46	−70.71	+	70.71
38	0.12	−5.97	−	5.97
39	−2.46	−20.20	+	20.20
40	1.78	7.80	+	7.80
41	−0.86	−8.91	+	8.91
42	−8.08	−55.53	+	55.53
43	−1.54	−15.58	+	15.58
44	−0.98	−10.86	+	10.86
45	−18.36	−117.48	+	117.48
46	−3.94	−25.17	+	25.17
47	1.18	6.25	+	6.25

and

$$w_5 = [-284.9 \quad 164.5 \quad 0.5233 \quad 14.59]\begin{bmatrix} 1 \\ 0.814 \\ 53 \\ 9.730 \end{bmatrix}$$

$$= 18.70$$

(The difference between 18.70 and 18.71 is due to round-off error.) We have chosen the initial vector of estimates to satisfy $y_i = \boldsymbol{b}'\boldsymbol{x}_i$ for $i = 1, 2, 3, 4$. This is why $z_1 = z_2 = z_3 = z_4 = 0$. The ith row of $\boldsymbol{A}$ is $\boldsymbol{x}_i'$ and the first column of $\boldsymbol{A}^{-1}$ is $\boldsymbol{d}_1$, so the $(i, 1)$ entry of the product $\boldsymbol{AA}^{-1}$ is $\boldsymbol{x}_i'\boldsymbol{d}_1$, which is w_i. Note that $\boldsymbol{AA}^{-1}$ is the identity matrix. This is why $w_1 = 1$ and $w_2 = w_3 = w_4 = 0$.

After calculating z_i and w_i, we determine the sign of z_i/w_i. If $w_i = 0$, this ratio is not defined, but this does not matter because such a data point does not contribute to the derivative since $|w_i| = 0$. Now we add up $|w_i|$ for all indices i where the sign of z_i/w_i is either negative or zero and subtract $|w_i|$ where the sign is positive. This yields -1221 as the derivative in the direction $\boldsymbol{d}_1$.

Having done the calculations for $\boldsymbol{d}_1$, there is a shortcut for calculating the derivative in the direction $-\boldsymbol{d}_1$. The table changes only slightly. The values of z_i and $|w_i|$ remain the same, but the signs of w_i change and hence the signs of z_i/w_i change except where $z_i = 0$. There are four data points with $z_i = 0$, one with $w_i = 1$ and the other three with $w_i = 0$. This implies that the derivative in the direction $-\boldsymbol{d}_1$ is $-(-1221 - 1) + 1 = 1223$. In general, the derivatives in the directions $\boldsymbol{d}$ and $-\boldsymbol{d}$ must add up to 2.

The derivatives for the four columns of $\boldsymbol{A}^{-1}$ and their negatives are -1221, 1223, -1323, 1325, 654, -652, 1903, and -1901. The most negative of these is -1901 in the direction $-\boldsymbol{d}_4$. So we look for a better vector of the form $\boldsymbol{b} + t\boldsymbol{d}_4$ (where t is negative—but it is not necessary to specify the sign of t because the algorithm automatically finds the best t regardless of sign).

Let $z_i = y_i - \boldsymbol{b}'\boldsymbol{x}_i$ and $w_i = \boldsymbol{d}_4'\boldsymbol{x}_i$. The value of t that gives the best vector is the slope of the best line passing through $(0, 0)$ that fits the data (w_i, z_i), $i = 1, 2, \ldots, 47$, $i \neq 7, 24$. An algorithm for this is presented in Section 4.2. Applying this algorithm, we find that the best line passing through $(0, 0)$ also passes through (w_{14}, z_{14}). Therefore the fourth rows of $\boldsymbol{A}$ and $\boldsymbol{c}$ are replaced by the data for area 14. That is, at the next step we have

$$\boldsymbol{A} = \begin{bmatrix} 1 & 0.604 & 29 & 11.744 \\ 1 & 0.765 & 44 & 9.323 \\ 1 & 0.735 & 36 & 9.948 \\ 1 & 0.077 & 11 & 13.686 \end{bmatrix} \quad \text{and} \quad \boldsymbol{c} = \begin{bmatrix} 1.825 \\ 2.251 \\ 2.351 \\ 0.693 \end{bmatrix}$$

At each step, four data points determine the current regression equation. They are sometimes called the *basis*. At each step, one of the data points in the basis is replaced by a data point from outside the basis. We started with data points 1, 2, 3, 4 as the initial basis. At the first step, point 4 was replaced by point 14. As the algorithm proceeds, point 3 is replaced by point 29, 2 by 26, 1 by 23, 26 by 10, 29 by 39, 10 by 19, 14 by 45, and 23 by 37. After this last step, the basis consists of points 37, 19, 39, and 45. When the derivatives for the eight directions are calculated at this step, we find that they are all positive. So the algorithm stops. The LAD regression equation is determined by the data points 37, 19, 39, 45:

$$\hat{\boldsymbol{\beta}} = \begin{bmatrix} 1 & 0.020 & 17 & 13.842 \\ 1 & 0.842 & 29 & 11.084 \\ 1 & 0.559 & 42 & 10.332 \\ 1 & 0.114 & 3 & 10.080 \end{bmatrix}^{-1} \begin{bmatrix} 1.224 \\ 1.974 \\ 2.351 \\ 1.946 \end{bmatrix} = \begin{bmatrix} 4.362 \\ -0.09098 \\ 0.01299 \\ -0.2425 \end{bmatrix}$$

$$\hat{Y} = 4.362 - 0.09098X_1 + 0.01299X_2 - 0.2425X_3$$

Nonuniqueness and Degeneracy. LAD multiple regression uses the LAD simple regression procedure to find the best vector of estimates in a given direction. Section 4.3 describes how nonuniqueness and degeneracy can affect this procedure. Moreover, similar problems can arise when more than one direction vector has the most negative derivative or when the least derivative is 0.

Regardless of any nonuniqueness or degeneracy that may occur during the algorithm, if the algorithm arrives at a step where the derivatives are positive in all $2(p + 1)$ directions, then the current vector of estimates is the unique LAD vector of estimates. If the algorithm arrives at a step where the derivatives are either positive or zero in all directions, then the current vector of estimates is an LAD vector and other LAD vectors can be found in the directions with derivative 0.

Sometimes the algorithm will iterate in circles without ever arriving at a step where the derivatives are all nonnegative. This is called cycling. There are several possible remedies. Theoretically, since the LAD vector of estimates (or at least one of them, in case of nonuniqueness) is known to pass through at least $p + 1$ of the data points, one could form all possible subsets of $p + 1$ data points, determine the corresponding vector of estimates ($\boldsymbol{b} = \boldsymbol{A}^{-1}\boldsymbol{c}$, as described previously), evaluate the sum of absolute deviations (4.9) for each vector, and then choose the one (or ones) with the smallest sum. However, this would require an impossible amount of computing for very large data sets. For the fire data, with $n = 45$ and $p = 3$, there would be 148,995 subsets of data points to consider.

A more feasible strategy when cycling occurs is to rerun the algorithm with a different vector for the initial vector of estimates. Another strategy is to perturb the data, that is, generate some very small random numbers and add them to the explanatory variables. This should eliminate the cycling, and the $p + 1$ data points that determine the LAD vector for the perturbed data are likely also to determine the LAD vector for the original data. This can be checked by seeing whether the derivatives in all $2(p + 1)$ directions are nonnegative.

Degeneracy sometimes leads to a situation in which the minimizing value of t in (4.10) is $t^* = 0$. If so, one can try a different direction vector $\boldsymbol{d}$, provided the derivative in the new direction is also negative.

4.7 TESTING $\beta_{q+1} = \cdots = \beta_p = 0$

Consider the general linear regression model $Y = \beta_0 + \beta_1 X_1 + \cdots + \beta_p X_p + e$. The least-squares test statistic for testing $\beta_{q+1} = \cdots = \beta_p = 0$ is

$$F_{\text{LS}} = \frac{\text{SSR}_{\text{reduced}} - \text{SSR}_{\text{full}}}{(p - q)\hat{\sigma}^2}$$

where SSR stands for the sum of squares of the residuals, $\text{SSR} = \Sigma \hat{e}_i^2$. A similar test statistic is used in least-absolute-deviations regression:

$$F_{\text{LAD}} = \frac{\text{SAR}_{\text{reduced}} - \text{SAR}_{\text{full}}}{(p - q)(\hat{\tau}/2)} \tag{4.12}$$

where SAR stands for the sum of absolute values of the residuals, $\text{SAR} = \Sigma |\hat{e}_i|$. The estimate $\hat{\tau}$ is given by (4.5) with $m = n - (p + 1)$.

When the random errors are assumed to have a normal distribution and when the null hypothesis is true, the test statistic F_{LS} is known to have an F distribution, regardless of the size n of the sample. When the distribution of the random errors is not specified, then F_{LS} has approximately an F distribution when the null hypothesis is true, provided the sample size n is large. This also holds for the test statistic F_{LAD}. Thus, for large n, we can calculate the approximate p-value of the test to be $\text{Prob}[F \geq F_{\text{LAD}}]$, where F denotes a random variable having an F distribution with $p - q$ and $n - p - 1$ degrees of freedom.

The approximation of the p-value is improved by the following modification. Calculate the p-value to be $\text{Prob}[G \geq (p - q)(1 - (p - q)/n)F_{\text{LAD}}]$, where G denotes a random variable having a chi-squared distribution with

Table 4.5 Residuals from the LAD Regression Analysis of the Fire Data, in Increasing Order

−0.6652	−0.1361	0.2488
−0.5547	−0.1012	0.2850
−0.5261	−0.0356	0.3277
−0.5202	−0.0188	0.3347
−0.4857	−0.0106	0.3439
−0.4790	0.0000	0.3604
−0.4647	0.0000	0.4457
−0.4233	0.0000	0.4642
−0.3914	0.0000	0.5815
−0.3520	0.0008	0.5939
−0.3219	0.1006	0.7547
−0.2786	0.1155	0.8047
−0.2323	0.1563	0.8355
−0.1565	0.2052	1.0740
−0.1492	0.2468	1.2015

$p - q$ degrees of freedom. For very large n, this is almost the same as $\text{Prob}[F \geq F_{\text{LAD}}]$ because $1 - (p - q)/n \approx 1$ and $(p - q)F$ with $n = \infty$ has the same distribution as G. For moderately large n, the p-value based on G has been found to be more accurate.

The Fire Data. We can apply the test above, with $p = 3$ and $q = 0$, to the fire data to test $\beta_1 = \beta_2 = \beta_3 = 0$, that is, to test whether any of the variables AGE, THEFT, and INCOME have a significant relationship with FIRE.

We want to calculate F_{LAD} in (4.12). For this we need the residuals $\hat{e}_i = y_i - (4.362 - 0.09098x_{i1} + 0.01299x_{i2} - 0.2425x_{i3})$ listed in Table 4.5. The sum of their absolute values is $\text{SAR}_{\text{full}} = 15.78$. The $m = 41$ nonzero residuals are used to calculate $\hat{\tau}$. Since $(41 + 1)/2 - \sqrt{41} = 14.60$, we have $k_1 = 15$. Similarly, $k_2 = 27$. From Table 4.5 we see $\hat{e}_{(15)} = -0.1492$ and $\hat{e}_{(27)} = 0.2488$. Hence $\hat{\tau} = \sqrt{41}(0.2488 + 0.1492)/4 = 0.6371$. The reduced model is $Y = \beta_0 + e$. The LAD estimate of β_0 in this model is simply the sample median $\tilde{y}$, which for the fire data is 2.251. So $\text{SAR}_{\text{reduced}} = \Sigma|y_i - 2.251| = 27.29$. Now $F_{\text{LAD}} = (27.29 - 15.78)/3(0.6371/2) = 12.04$.

To get the p-value, we first calculate $(p - q)(1 - (p - q)/n)F_{\text{LAD}} = 3(1 - 3/45)(12.04) = 33.71$. The p-value is approximately $\text{Prob}[G \geq 33.71]$, where G has a chi-squared distribution with 3 degrees of freedom. The chi-squared table in the Appendix tells us that $\text{Prob}[G \geq 16.29] = 0.001$. So the p-value is less than 0.001. We conclude that at least some of the explanatory variables have a significant relationship with FIRE.

4.8 COMPUTATION

Calculation of LAD estimates for a simple linear regression model by the method described in Section 4.2 involves the construction of a series of tables like Table 4.1. Using only a hand calculator and paper and pencil, this is somewhat tedious even for small data sets. If you correctly guess one of the points on the LAD regression line, only two tables are required.

LAD regression estimates are obtainable from the function *l1fit* in the computer language S-PLUS and from the robust regression package ROBSYS. A short FORTRAN program could be written to calculate LAD estimates using the IMSL subroutine RLLAV. If the IMSL library is not available to you, you could copy the FORTRAN program for simple LAD regression published by Sadovski (1974), the PASCAL program for simple LAD regression by Farebrother (1988), or the FORTRAN program for multiple LAD regression by Barrodale and Roberts (1974).

Minimization of (4.9) can be formulated as a linear programming (LP) problem, as in Arthanari and Dodge (1981, Section 2.7) or Bloomfield and Steiger (1983, Section 6.2). Therefore one could use a linear programming computer package to obtain LAD estimates. For large sample sizes, it is easier to solve the dual LP problem.

If you are able to calculate the LAD regression estimates, then you can also calculate the test statistic F_{LAD} in (4.12). Using the regression estimates, form the residuals, from which you can obtain $\hat{\tau}$ and SAR_{full}. Similarly, you can obtain $\text{SAR}_{\text{reduced}}$. Now F_{LAD} can be calculated. If $q = 0$, then $\text{SAR}_{\text{reduced}} = \Sigma|y_i - \tilde{y}|$, where $\tilde{y}$ is the sample median of the y_i's.

Test Case. To try out a computational procedure for least-absolute-deviations regression, the data set in Table 3.2 can serve as a test case. For simple regression, let $X = X_1$. The estimated regression line is $\hat{Y} = 30.00 + 1.750X$. The test statistic for testing $\beta = 0$ is $|t| = 2.676$. For multiple regression, using both X_1 and X_2, the estimated regression equation is $\hat{Y} = 32.71 + 1.595X_1 - 0.09524X_2$. The test statistic for testing $\beta_1 = \beta_2 = 0$ is $F = 4.900$.

NOTES

4.1. Boscovich (1711–1787) was born in Ragusa (now called Dubrovnik), became a Jesuit priest, and spent most of his life in Rome. He was a prominent figure in eighteenth century European science. In 1757 he formulated the principle of least absolute deviations, and three years later he published a detailed procedure for implementing the principle. Laplace

(1749–1827) is regarded as the greatest French mathematician of his time. He played a central role in the early development of probability and statistics.

4.2a. LAD regression is sometimes called L_1-regression because (4.1) is the L_1-norm of the vector of deviations. The L_1-norm of a vector $\boldsymbol{v}$ is $\Sigma|v_i|$. Similarly, least-squares regression can be called L_2-regression because it minimizes the L_2-norm of the vector of deviations. The L_2-norm of a vector $\boldsymbol{v}$ is $\sqrt{\Sigma v_i^2}$.

4.2b. Conditions (4.2) can be derived by differentiating function (4.4). This function is differentiable at all values of b except the n ratios $b_i = (y_i - y_0)/(x_i - x_0)$. The derivative of (4.4) in the interval $b_{i-1} < b < b_i$ is $|x_1 - x_0| + \cdots + |x_{i-1} - x_0| - |x_i - x_0| - \cdots - |x_n - x_0|$. This can be seen as follows. Let d_i denote the deviation $(y_i - y_0) - b(x_i - x_0)$. Then (4.4) is $\Sigma|d_i|$. We can write $|d_i| = |(y_i - y_0) - b(x_i - x_0)| = |x_i - x_0|\,|(y_i - y_0)/(x_i - x_0) - b| = |x_i - x_0|\,|b_i - b|$. If $b < b_i \leq \cdots \leq b_n$, then $|d_i| = |x_i - x_0|(b_i - b), \ldots, |d_n| = |x_n - x_0|(b_n - b)$. If $b_1 \leq \cdots \leq b_{i-1} < b$, then $|d_1| = |x_1 - x_0|(b - b_1), \ldots, |d_{i-1}| = |x_{i-1} - x_0|(b - b_{i-1})$. Therefore, if $b_{i-1} < b < b_i$, then (4.4) is equal to

$$b\big(|x_1 - x_0| + \cdots + |x_{i-1} - x_0| - |x_i - x_0| - \cdots - |x_n - x_0|\big) + c$$

where c is a constant not involving b. The derivative of this linear function of b is simply the coefficient of b.

Thus the derivative is $R_i - S_i$, where $R_i = |x_1 - x_0| + \cdots + |x_{i-1} - x_0|$ and $S_i = |x_i - x_0| + \cdots + |x_n - x_0|$. Since $R_i + S_i = T$, the derivative can be written as $R_i - (T - R_i) = 2R_i - T$. Therefore the derivative is negative if and only if $R_i < \frac{1}{2}T$. So the first condition in (4.2) says that the derivative of (4.4) is negative for $b < b_k$, and the second condition says that the derivative is positive for $b > b_k$.

4.2c. After estimating the regression coefficients, we should calculate residuals and use them to check the validity of the model as described in Chapter 2. To detect outliers it helps to standardize the residuals by dividing them by an estimate of σ, the standard deviation of the population of errors. In a least-squares analysis we divide the residuals by $\hat{\sigma}$ in (3.3). But the estimate $\hat{\sigma}$ is itself sensitive to outliers. A less sensitive estimate is $\tilde{\sigma} = 1.483\text{MAD}$, where MAD is the median of the absolute values of the nonzero residuals.

Sometimes MAD is computed using all the residuals (including the zeros). This option is available in the robust regression computer package ROBSYS (Marazzi, 1987, p. 148). Hill and Holland (1977) and McKean and Schrader (1987) have found that estimates based on all the residuals tend to be too small. To obtain a larger estimate, they suggest using only the nonzero residuals.

The multiplier 1.483 ensures that, if the population of errors is assumed to have a normal distribution, then $\tilde{\sigma}$ is a "consistent" estimate of σ. Consistency means that, regarding $\tilde{\sigma}$ as a random variable, for very large samples there is a very high probability that $\tilde{\sigma}$ will be very close to σ.

4.3. The LAD regression line is not necessarily unique. For an illustration of nonuniqueness, consider the data set consisting of the following three data points: (1, 1), (2, 1), (2, 2). Make a scatter plot of these data. From the plot you can see that if we want a line that minimizes the sum of absolute deviations, the line should pass through (1, 1) and pass between the two points (2, 1) and (2, 2). In other words, the minimum of the sum

$$|1 - (a + b)| + |1 - (a + 2b)| + |2 - (a + 2b)|$$

is achieved by setting $\hat{\alpha} + \hat{\beta} = 1$ and $\hat{\alpha} + 2\hat{\beta} = t$, where t can be any number in the interval $1 \le t \le 2$. Thus $0 \le \hat{\beta} \le 1$ and $\hat{\alpha} = 1 - \hat{\beta}$. The nonuniqueness occurs because $|1 - t| + |2 - t|$ is minimized not just at a single value of t but at any value in the interval $1 \le t \le 2$.

In contrast, the LS regression line is unique. The minimum of the sum

$$[1 - (a + b)]^2 + [1 - (a + 2b)]^2 + [2 - (a + 2b)]^2$$

is achieved by setting $\hat{\alpha} + \hat{\beta} = 1$ and $\hat{\alpha} + 2\hat{\beta} = 1.5$. Thus $\hat{\alpha} = 0.5$ and $\hat{\beta} = 0.5$. Note that $(1 - t)^2 + (2 - t)^2$ is minimized uniquely at $t = 1.5$.

4.4a. Formula (4.5) for estimating τ is recommended by McKean and Schrader (1987, p. 301).

4.4b. Suppose $e_1, e_2, \ldots, e_n$ are a random sample of n independent observations from a population with median ν. Let us show that the interval from $e_{(k_1)}$ to $e_{(k_2)}$ is an approximate 95%-confidence interval for ν, where k_1 and k_2 are the closest integers to $(n + 1)/2 \pm \sqrt{n}$. Let $p = \text{Prob}[e_{(k_1)} < \nu < e_{(k_2)}]$. Regarding the e_i's as random variables, we must show that $p \approx 0.95$. Note that the event $e_{(k)} < \nu$ is equivalent to the event that at least k of the errors e_i are less than ν. Therefore, letting x be the number of errors that are less than ν, we can write $p = \text{Prob}[k_1 \le x < k_2]$. To adjust for the discreteness of x, we write this probability as $\text{Prob}[k_1 - \frac{1}{2} < x < k_2 - \frac{1}{2}]$.

Now we use the fact that x has a binomial distribution and that, for large n, a binomial distribution is approximately normal. The distribution of x is binomial because x is the number of successes in n independent trials, where the ith trial consists of observing whether e_i is less than ν. The probability of success on a single trial is $\frac{1}{2}$ since, by the definition of the median, each observation has a $\frac{1}{2}$ chance of being less than ν. So the mean and standard deviation of x are $n(\frac{1}{2}) = n/2$ and $\sqrt{n(\frac{1}{2})(1-\frac{1}{2})} = \sqrt{n}/2$. Hence $0.95 \approx \text{Prob}[n/2 - 2(\sqrt{n}/2) < x < n/2 + 2(\sqrt{n}/2)] = \text{Prob}[k_1 - \frac{1}{2} < x < k_2 - \frac{1}{2}] = p$.

Note that the two confidence intervals that we have compared in order to justify formula (4.5) are both based on normal approximations. To construct the interval from $\tilde{e} - 2\hat{\tau}/\sqrt{n}$ to $\tilde{e} + 2\hat{\tau}/\sqrt{n}$, we approximated the distribution of $\tilde{e}$ by a normal distribution. To construct the interval from $e_{(k_1)}$ to $e_{(k_2)}$, we approximated the distribution of x by a normal distribution.

4.4c. Theory tells us that, for infinitely large sample sizes, the p-value of the LAD test in Section 4.4 can be obtained from the standard normal distribution (see Koenker and Bassett (1982) and Note 4.4d). If the standard normal distribution is used for sample sizes as small as $n = 20$, Dielman and Pfaffenberger (1990) have found that the p-values tend to be smaller than they should be. To compensate for this, we have used the t-distribution instead.

4.4d. Koenker and Bassett (1982) presented three types of LAD tests: Wald, likelihood ratio, and Lagrange multiplier tests. The test statistic in Section 4.4 is the square root of a Wald test statistic. The likelihood ratio test is the one described in Section 4.7. The Lagrange multiplier test statistic for testing $\beta = 0$ in simple regression is equal to n times the squared correlation coefficient between the x_i's and the u_i's, where u_i is the sign of $y_i - \tilde{y}$ and $\tilde{y}$ is the median of the y_i's. The p-value is calculated by assuming the test statistic has a chi-squared distribution with 1 degree of freedom. See Note 4.7a.

4.6a. The algorithm in Section 4.6 is a version of an algorithm by Barrodale and Roberts (1974). It is based on the simplex algorithm for solving linear programming problems. See Bloomfield and Steiger (1983, Section 7.2.1) and Gentle, Narula, and Sposito (1987). For the sake of simplicity, our algorithm arbitrarily picks the first $p + 1$ data points to form an initial vector of estimates, whereas the Barrodale-Roberts algorithm includes a special start-up phase.

4.6b. Let us calculate the right-hand derivative of (4.11) at $t = 0$. Reindex the data points so that the ratios z_i/w_i are in nondecreasing order. As in Note 4.2b, the derivative of (4.11) is $|w_1| + \cdots + |w_{i-1}| - |w_i| - \cdots - |w_n|$ for $z_{i-1}/w_{i-1} < t < z_i/w_i$. So the right-hand derivative at $t = 0$ is $|w_1| + \cdots + |w_{k-1}| - |w_k| - \cdots - |w_n|$, where k is the smallest integer such that $z_k/w_k > 0$. This can be expressed as $W_- + W_0 - W_+$.

4.6c. The $p + 1$ data points that are chosen at the start of the algorithm to form the matrix $A = (\boldsymbol{x}_1, \boldsymbol{x}_2, \ldots, \boldsymbol{x}_{p+1})'$ must be chosen so that A is invertible. For observational data, such as the fire data, this is rarely a problem. But in a designed experiment, where the values of the explanatory variables are chosen according to some pattern, care must be taken to choose data points such that none of the $p + 1$ vectors $\boldsymbol{x}_i$ is a linear combination of the others.

4.6d. Let us verify that $\boldsymbol{b} + t^*\boldsymbol{d}_3 = \boldsymbol{A}^{*-1}\boldsymbol{c}^*$, as stated in the subsection on justification of the algorithm. We want to show that $\boldsymbol{A}^*\boldsymbol{b} + t^*\boldsymbol{A}^*\boldsymbol{d}_3 = \boldsymbol{c}^*$. The components of this vector equation are $\boldsymbol{x}_i'\boldsymbol{b} + t^*\boldsymbol{x}_i'\boldsymbol{d}_3 = y_i$ $(i = 1, 2, 4)$ and $\boldsymbol{x}_k'\boldsymbol{b} + t^*\boldsymbol{x}_k'\boldsymbol{d}_3 = y_k$. The last equation follows from the definition of t^*. From the fact that $\boldsymbol{A}\boldsymbol{A}^{-1}$ is the identity matrix, we see that $\boldsymbol{x}_i'\boldsymbol{d}_3 = 0$ for $i = 1, 2, 4$. Now note that $\boldsymbol{x}_i'\boldsymbol{b} = y_i$ because $\boldsymbol{A}\boldsymbol{b} = \boldsymbol{c}$.

4.6e. The vector $\hat{\boldsymbol{\beta}}_{\text{LAD}}$ of least-absolute-deviations estimates has a variance–covariance matrix approximately equal to $\tau^2(\boldsymbol{X}'\boldsymbol{X})^{-1}$, where τ is described in Section 4.4 and $\boldsymbol{X}$ is the matrix of explanatory variables with a vector of 1's as the first column. Similarly, as seen in Note 3.7, the vector $\hat{\boldsymbol{\beta}}_{\text{LS}}$ of least-squares estimates has variance–covariance matrix $\sigma^2(\boldsymbol{X}'\boldsymbol{X})^{-1}$. Therefore the standard deviation of $\hat{\beta}_{j,\text{LAD}}$ is approximately τ/σ times the standard deviation of $\hat{\beta}_{j,\text{LS}}$.

4.7a. Three different types of LAD tests were proposed by Koenker and Bassett (1982). The test in Section 4.7 is their likelihood ratio test. To describe the other two tests, partition the parameter vector into two parts $\boldsymbol{\beta}_1 = (\beta_0, \beta_1, \ldots, \beta_q)$ and $\boldsymbol{\beta}_2 = (\beta_{q+1}, \ldots, \beta_p)$, and let $\boldsymbol{X}_1$ and $\boldsymbol{X}_2$ be the matrices of corresponding explanatory variables. We want to test whether $\boldsymbol{\beta}_2$ is the zero vector. Let $\hat{\boldsymbol{\beta}}_2$ be its LAD estimate in the full model. The Wald test statistic is $\hat{\boldsymbol{\beta}}_2'\boldsymbol{V}_2^{-1}\hat{\boldsymbol{\beta}}_2$, where $\boldsymbol{V}_2$ is an estimate of $\text{Cov}(\hat{\boldsymbol{\beta}}_2)$, namely $\hat{\tau}^2\boldsymbol{W}$, where $\boldsymbol{W} = (\boldsymbol{X}_2'\boldsymbol{X}_2 - \boldsymbol{X}_2'\boldsymbol{X}_1(\boldsymbol{X}_1'\boldsymbol{X}_1)^{-1}\boldsymbol{X}_1'\boldsymbol{X}_2)^{-1}$. When $p = 1$ and $q = 0$, this is the square of the test statistic in Section 4.4. Let $\boldsymbol{u}$ be the vector of signs of the LAD residuals in the reduced model with $\boldsymbol{\beta}_2 = \boldsymbol{0}$. The Lagrange multiplier test statistic is $\boldsymbol{u}'\boldsymbol{X}_2\boldsymbol{W}\boldsymbol{X}_2'\boldsymbol{u}$. This test avoids the problem of estimating τ.

Koenker and Bassett (1982) showed that the three tests are asymptotically equivalent. For infinitely large samples, they have the same power and their p-values can be calculated by assuming the test statistic has a chi-squared distribution with $p - q$ degrees of freedom. Dielman and Pfaffenberger (1990) applied the three tests to simulated data following a simple linear regression model for sample sizes $n = 20$, 40 and 100. For $n = 20$, the Lagrange multiplier test was more powerful than the other two tests when the error distribution was heavy-tailed, but the likelihood ratio test had somewhat greater power when the error distribution was normal. For $n = 40$ and $n = 100$, the likelihood ratio test tended to have greater power.

4.7b. The modified calculation of the p-value based on G is recommended by Schrader and McKean (1987).

4.8. In S-PLUS, the function that performs LAD regression is *l1fit*. It uses the Barrodale–Roberts (1974) algorithm, as does ROBSYS and the IMSL subroutine RLLAV. This is basically the same algorithm as the one described in Section 4.6, but Barrodale and Roberts have devised refinements to speed up the algorithm.

Additional Reading. For more on theory and computation for LAD regression, see the book by Bloomfield and Steiger (1983) and the conference proceedings edited by Dodge (1987a, 1987b, 1992). For a review article, see Dielman and Pfaffenberger (1982).

REFERENCES

Andrews, D. F., and A. M. Herzberg (1985). *Data*. Springer-Velag, New York.

Arthanari, T. S., and Y. Dodge (1981). *Mathematical Programming in Statistics*. Wiley, New York.

Barrodale, I., and F. D. K. Roberts (1974). Algorithm 478: Solution of an overdetermined system of equations in the l_1-norm. *Communications of the Association for Computing Machinery*, vol. 17, pp. 319–320.

Bloomfield, P., and W. Steiger (1983). *Least Absolute Deviations: Theory, Applications, and Algorithms*. Birkhäuser, Boston.

Dielman, T., and R. Pfaffenberger (1982). LAV (least absolute value) estimation in linear regression: a review. In S. H. Zanakis and J. S. Rustagi (eds.), *Optimization in Statistics*. North-Holland, New York.

Dielman, T., and R. Pfaffenberger (1990). Tests of linear hypotheses and LAV estimation: a Monte Carlo comparison. *Communications in Statistics—Simulation and Computation*, vol. 19, pp. 1179–1199.

Dodge, Y. (ed.) (1987a). *Statistical Data Analysis Based on the L_1-Norm and Related Methods*. North-Holland, New York.

Dodge, Y. (ed.) (1987b). Special Issue on Statistical Data Analysis Based on the L_1 Norm and Related Methods. *Computational Statistics & Data Analysis*, vol. 5, pp. 237–450.

Dodge, Y. (ed.) (1992). *L_1-Statistical Analysis and Related Methods*. North-Holland, Amsterdam.

Farebrother, R. W. (1988). Algorithm AS 238: A simple recursive procedure for the L_1 norm fitting of a straight line. *Applied Statistics*, vol. 37, pp. 457–465.

Gentle, J. E., S. C. Narula, and V. A. Sposito (1987). Algorithms for unconstrained L_1 linear regression. In Y. Dodge (ed.) (1987a).

Hill, R. W., and P. W. Holland (1977). Two robust alternatives to least squares regression. *Journal of the American Statistical Association*, vol. 72, pp. 828–833.

Koenker, R., and G. Bassett (1982). Tests of linear hypotheses and l_1 estimation. *Econometrica*, vol. 50, pp. 1577–1583.

Marazzi, A. (1987). Solving bounded influence regression problems with ROBSYS. In Y. Dodge (ed.) (1987a).

McKean, J. W., and R. M. Schrader (1987). Least absolute errors analysis of variance. In Y. Dodge (ed.) (1987a).

Sadovski, A. N. (1974). Algorithm AS 74: L_1-norm fit of a straight line. *Applied Statistics*, vol. 23, pp. 244–248.

Schrader, R. M., and J. W. McKean (1987). Small sample properties of least absolute errors analysis of variance. In Y. Dodge (ed.) (1987a).

Stigler, S. M. (1986). *The History of Statistics: The Measurement of Uncertainty Before 1900*. Harvard University Press, Cambridge, MA.

Whyte, L. L. (ed.) (1961). *Roger Joseph Boscovich*. Allen & Unwin, London.

CHAPTER 5

M-Regression

... And so the talk goes on
Between the believers and those who are not.

MOULAVI RUMI: *Persian Soufi Poet (1207–1273)*

5.1 INTRODUCTION

M-regression and LAD regression are part of what is called robust statistics. A statistical procedure is regarded as "robust" if it performs reasonably well even when the assumptions of the statistical model are not true. If we assume that our data follow a normal linear regression model, then least-squares estimates and test perform quite well, but they are not robust when the assumption of normality for the population of random errors is invalid. M-regression was specifically developed to be robust with respect to this assumption. Peter Huber introduced the idea of M-estimation in 1964.

5.2 AN EXAMPLE OF SIMPLE REGRESSION

Data from a study on the shelf life of packaged foods are shown in Table 5.1. Moisture content is the primary factor that determines shelf life, that is, the length of time during which the cereal is acceptable for eating. The higher the moisture content, the soggier the cereal, and hence the lower its acceptability. A particular brand of dry cereal was stored on a shelf at 73°F and 50% relative humidity. At various times during storage, the moisture content of the cereal was measured. We can use these data to see how moisture content depends on the time on the shelf. Time is measured in days.

First we make a plot of the data with x = days on the shelf and y = moisture content. This is shown in Figure 5.1. We see that a straight line

Table 5.1 Shelf Life Data

Identification Number of the Measurement	Moisture Content (Y)	Days on the Shelf (X)
1	2.8	0
2	3.0	3
3	3.1	6
4	3.2	8
5	3.4	10
6	3.4	13
7	3.5	16
8	3.1	20
9	3.8	24
10	4.0	27
11	4.1	30
12	4.3	34
13	4.4	37
14	4.9	41

Source: Devore and Peck (1986, pp. 471–472).

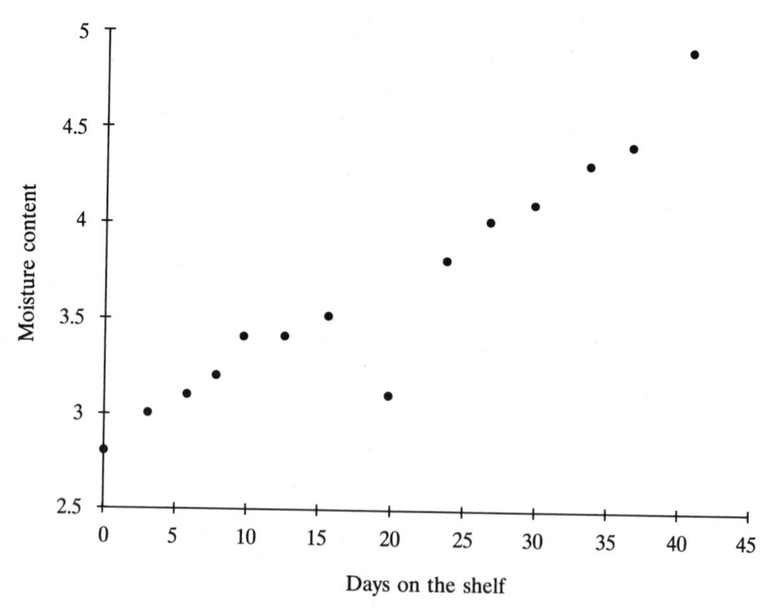

Figure 5.1 Plot of the shelf life data.

might describe the overall pattern of the data reasonably well, and so the simple linear regression model $Y = \alpha + \beta X + e$ seems like a good model to try.

5.3 ESTIMATING THE REGRESSION LINE

In least-squares estimation, $\hat{\alpha}$ and $\hat{\beta}$ are chosen so that $\Sigma \hat{e}_i^2$ is as small as possible. In least-absolute-deviations estimation, they are chosen so that $\Sigma |\hat{e}_i|$ is as small as possible. In M-estimation, this idea is generalized and $\hat{\alpha}$ and $\hat{\beta}$ are chosen so that $\Sigma \rho(\hat{e}_i)$ is as small as possible, where $\rho(e)$ is some function of e. Least-squares and least-absolute-deviations estimation can be regarded as the particular cases of M-estimation in which $\rho(e) = e^2$ and $\rho(e) = |e|$.

Definition of Huber M-Estimates. The M-estimates introduced in this chapter, called Huber M-estimates, use a function $\rho(e)$ that is a compromise between e^2 and $|e|$. The main advantage of LAD estimates over LS estimates is that they are not so sensitive to outliers. When there are no outliers, however, LS estimates may be more accurate. We can try to combine the advantages of both methods by defining $\rho(e)$ to be equal to e^2 when e is near 0 and equal to $|e|$ (or at least similar to $|e|$) when e is far from 0. Specifically, we define

$$\rho(e) = \begin{cases} e^2 & \text{if } -k \le e \le k \\ 2k|e| - k^2 & \text{if } e < -k \text{ or } k < e \end{cases} \tag{5.1}$$

Following a suggestion of Huber we take $k = 1.5\hat{\sigma}$, where $\hat{\sigma}$ is an estimate of the standard deviation σ of the population of random errors. In order to make $\rho(e)$ a smooth function, $2k|e| - k^2$ is used instead of $|e|$. A graph of this function is shown in Figure 5.2.

To estimate σ we use $\hat{\sigma} = 1.483$ MAD, where MAD is the median of the absolute deviations $|\hat{e}_i|$. The multiplier 1.483 is chosen to ensure that $\hat{\sigma}$ would be a good estimate of σ if it were the case that the distribution of the random errors were normal.

The *Huber M-estimates* $\hat{\alpha}$ and $\hat{\beta}$ are the values of a and b that minimize

$$\sum \rho(y_i - (a + bx_i)) \tag{5.2}$$

Note that a and b, besides appearing explicitly in (5.2) in the argument of the function ρ, also appear implicitly in the definition of ρ; the function ρ involves $k = 1.5\hat{\sigma}$ and $\hat{\sigma}$ is calculated from the deviations $y_i - (a + bx_i)$. Next we need an algorithm for minimizing (5.2).

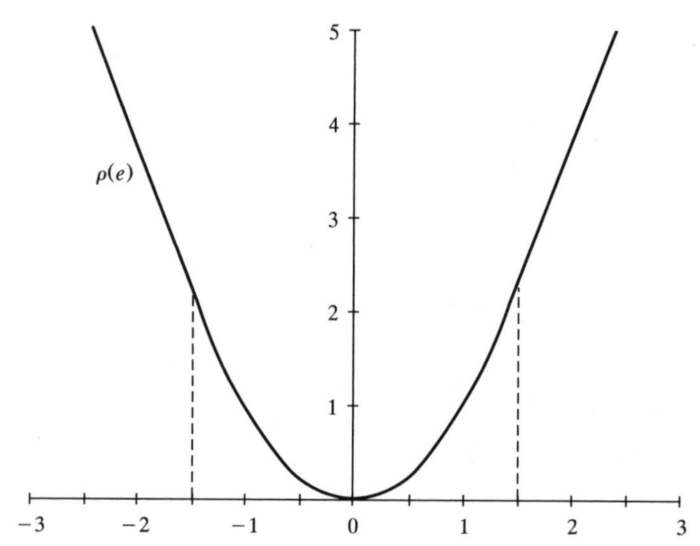

Figure 5.2 Graph of the function $\rho(e)$ used in the definition of Huber M-estimates, $k = 1.5$.

An Algorithm. To begin the algorithm, for initial estimates of α and β let us take the least-squares estimates. These are used to calculate deviations and an estimate of σ. These in turn are used, as described in the next paragraph, to obtain improved estimates of α and β. These improved estimates are used to calculate new deviations and an improved estimate of σ. Then the new deviations and new $\hat{\sigma}$ are used to obtain improved estimates of α and β. The algorithm is iterated in this way until a step is reached at which the improved estimates are the same (or at least approximately the same) as the previous estimates.

To be more specific, at any step in the algorithm, let a^0 and b^0 be the current estimates of α and β. Calculate the deviations $y_i - (a^0 + b^0 x_i)$ and from them calculate $\hat{\sigma}^0 = 1.483\text{MAD}$. Now make the following adjustment of the y-values to get rid of large deviations. The deviation of y_i from the current estimated regression line is $e_i^0 = y_i - (a^0 + b^0 x_i)$. Thus $y_i = a^0 + b^0 x_i + e_i^0$. Now define $y_i^* = a^0 + b^0 x_i + e_i^*$, where e_i^* is the adjusted deviation obtained by truncating e_i^0 so that none of the deviations is larger than $1.5\hat{\sigma}^0$ in absolute value. That is, $e_i^* = e_i^0$ (and hence $y_i^* = y_i$) if e_i^0 is between $-1.5\hat{\sigma}^0$ and $1.5\hat{\sigma}^0$, $e_i^* = -1.5\hat{\sigma}^0$ if e_i^0 is less than $-1.5\hat{\sigma}^0$, and $e_i^* = 1.5\hat{\sigma}^0$ if e_i^0 is greater than $1.5\hat{\sigma}^0$. Let the improved estimates of α and β be the least-squares estimates obtained from the adjusted data $y_1^*, \ldots, y_n^*$.

Justification of the Algorithm. Even though the algorithm seems reasonable, it may not be clear how in minimizes (5.2). With $\hat{\sigma}$ kept fixed, we can minimize (5.2) by taking derivatives with respect to a and b and setting them equal to 0. This yields two equations in the two unknowns a and b:

$$\sum \rho'(y_i - (a + bx_i)) = 0$$
$$\sum x_i \rho'(y_i - (a + bx_i)) = 0 \qquad (5.3)$$

Note that the derivative $\rho'(e)$ has the same value $-3\hat{\sigma}$ for all e less than or equal to $-1.5\hat{\sigma}$, and $\rho'(e)$ has the same value $3\hat{\sigma}$ for all e greater than or equal to $1.5\hat{\sigma}$. Therefore the solutions to (5.3) remain the same if the deviations $e_i = y_i - (a + bx_i)$ are replaced by the truncated deviations e_i^*, where $e_i^* = e_i$ if e_i is between $-1.5\hat{\sigma}$ and $1.5\hat{\sigma}$, $e_i^* = -1.5\hat{\sigma}$ if e_i is less than $-1.5\hat{\sigma}$, and $e_i^* = 1.5\hat{\sigma}$ if e_i is greater than $1.5\hat{\sigma}$. That is, the y_i can be replaced by the adjusted values $y_i^* = a_i + bx_i + e_i^*$ without changing the solutions to (5.3), hence without changing the values of a and b that minimize (5.2). As a result of the adjustment, $\rho(y_i^* - (a_i + bx_i)) = [y_i^* - (a + bx_i)]^2$. Minimizing $\Sigma[y_i^* - (a_i + bx_i)]^2$ yields, by definition, the least-squares estimates obtained from the adjusted data. In the algorithm, y_i^* is adjusted relative to the current estimate of the regression line rather than the final M-estimate, but this discrepancy disappears when the algorithm converges.

The Shelf Life Data. To begin the algorithm for finding M-estimates of α and β, we take for initial estimates the least-squares estimates, which are $a^0 = 2.786$ and $b^0 = 0.04662$. Calculate the fitted y-values $\hat{y}_i^0 = a^0 + b^0 x_i$ and then the deviations $e_i^0 = y_i - \hat{y}_i^0$, as in Table 5.2. The median of the absolute deviations $|e_i^0|$ is MAD = 0.04161. So $\hat{\sigma}^0 = (1.483)(0.04161) = 0.06171$ and $1.5\hat{\sigma}^0 = 0.09257$. Now truncate the deviations e_i^0 to obtain e_i^*. Two of the deviations are greater than $1.5\hat{\sigma}^0$, namely, $e_5^0 = 0.16829$ and $e_{14}^0 = 0.28506$. These are truncated to $e_5^* = e_{14}^* = 0.09257$. One deviation is less than $-1.5\hat{\sigma}^0$, namely, $e_8^0 = -0.57792$. This is truncated to $e_8^* = -0.09257$. Now the adjusted y-values are obtained as $y_i^* = \hat{y}_i^0 + e_i^*$. Note that $y_i^* = y_i$ for all i except $i = 5$, 8, and 14.

The initial estimates 2.786 and 0.04662 were obtained by applying the least-squares method to the data in Table 5.1. If we replace the y-values in that table by the adjusted y-values for the last column of Table 5.2 and we then apply the least-squares method of estimation, we obtain new estimates $a^0 = 2.827$ and $b^0 = 0.04325$. This completes the first iteration of the algorithm.

For the second iteration we construct a new table similar to Table 5.2. The first column remains the same. The new estimates a^0 and b^0 are used to calculate new fitted y-values $\hat{y}_i^0$ for column 2. After the last column of the new table is calculated, the least-squares method is applied to the new

Table 5.2 Calculations in the First Iteration of the Algorithm for Obtaining M-Estimates for the Shelf Life Data

Observed y-Value y_i	Fitted y-Value $\hat{y}_i^0$	Deviation e_i^0	Truncated Deviation e_i^*	Adjusted y-Value y_i^*
2.8	2.78551	0.01449	0.01449	2.80000
3.0	2.91937	0.08063	0.08063	3.00000
3.1	3.05323	0.04677	0.04677	3.10000
3.2	3.14247	0.05753	0.05753	3.20000
3.4	3.23171	0.16829	0.09257	3.32428
3.4	3.36557	0.03443	0.03443	3.40000
3.5	3.49944	0.00056	0.00056	3.50000
3.1	3.67792	−0.57792	−0.09257	3.58535
3.8	3.85640	−0.05640	−0.05640	3.80000
4.0	3.99026	0.00974	0.00974	4.00000
4.1	4.12412	−0.02412	−0.02412	4.10000
4.3	4.30260	−0.00260	−0.00260	4.30000
4.4	4.43646	−0.03646	−0.03646	4.40000
4.9	4.61494	0.28506	0.09257	4.70751

adjusted y-values to obtain updated estimates $a^0 = 2.835$ and $b^0 = 0.04285$. Iteration is continued until the estimates converge. After seven iterations, the relative difference between the updated estimate and the previous estimate is less than 10^{-4} for both α and β. Thus, to an accuracy of about four significant digits, we obtain $\hat{\alpha} = 2.838$ and $\hat{\beta} = 0.04264$.

5.4 TESTING $\beta = 0$

Section 5.7 describes how to test $\beta_{q+1} = \cdots = \beta_p = 0$ in a multiple linear regression model. Testing $\beta = 0$ is the special case in which $p = 1$ and $q = 0$. The test statistic is F_M in formula (5.6). An approximate p-value can be calculated as $\text{Prob}[F \geq F_M]$, where F denotes a random variable having an F distribution with 1 and $n - 2$ degrees of freedom, or as $\text{Prob}[|t| \geq |t_M|]$, where $|t_M| = \sqrt{F_M}$ and t denotes a random variable having a t distribution with $n - 2$ degrees of freedom.

The Shelf Life Data. To calculate F_M we first calculate the residuals $\hat{e}_i = y_i - (\hat{\alpha} + \hat{\beta}x_i)$ using the M-estimates $\hat{\alpha} = 2.838$ and $\hat{\beta} = 0.04264$ obtained in Section 5.3. The median of the absolute residuals $|\hat{e}_i|$ is MAD = 0.02056, so $\hat{\sigma} = (1.483)(0.2056) = 0.03049$ and $1.5\hat{\sigma} = (1.5)(0.03049) = 0.04574$. For the 10 residuals between -0.04574 and 0.04574, $\rho(\hat{e}_i) = \hat{e}_i^2$, and for the other four residuals, $\rho(\hat{e}_i) = 0.09148|\hat{e}_i| - 0.002092$. Adding all 14 values of $\rho(\hat{e}_i)$, we obtain $\text{STR}_{\text{full}} = 0.09676$. By truncating the two residuals that are below -0.04574 and the two residuals that are above 0.04574, we obtain the truncated residuals $\hat{e}_i^*$. The sum of their squares is 0.01272. Now we can calculate $\hat{\lambda} = (14/10)(0.01272)/12 = 0.001485$.

The reduced model with $\beta = 0$ is $Y = \alpha + e$. In applying the Huber M-estimation procedure to this model, we modify the procedure in the following way. The estimate $\hat{\sigma}$ is not iterated. Throughout the iterations required to obtain the M-estimate $\hat{\sigma}$ in the reduced model, we keep the same estimate $\hat{\sigma} = 0.03049$, the estimate calculated from the *full* model. The procedure finds a value of a that minimizes the function $\Sigma\rho(y_i - a)$, where ρ is defined as in (5.1) with $k = 1.5\hat{\sigma} = (1.5)(0.03049) = 0.04574$. We start by using the least-squares estimate as an initial estimate of α. For the model $Y = \alpha + e$, the least-squares estimate of α is simply the sample mean $\bar{y}$, so $a^0 = \bar{y} = 3.643$. Now we calculate the deviations $e_i^0 = y_i - 3.643$ and truncate the deviations with absolute values greater than $1.5\hat{\sigma} = 0.04574$. Eight of the deviations e_i^0 are less than -0.04574 and are truncated to $e_i^* = -0.04574$; the other six deviations e_i^0 are greater than 0.04574 and are truncated to 0.04574. So eight of the adjusted y-values are $y_i^* = a^0 + e_i^* = 3.643 + (-0.04574) = 3.597$ and the other six are $y_i^* = a^0 + e_i^* = 3.643 + 0.04574 = 3.689$. The updated estimate of α is the sample mean of the adjusted y-values, which is 3.636. For the next iteration, we calculate the deviations $e_i^0 = y_i - 3.636$ and truncate the deviations with absolute values greater than 0.04574. After many iterations, the estimates converge to $\hat{\alpha} = 3.454$.

Now calculate the residuals $\hat{e}_i = y_i - 3.454$. All the residuals have absolute value greater than or equal to 0.04574, so $\rho(\hat{e}_i) = 0.09148\ |\hat{e}_i| - 0.002092$ for all i. Adding the 14 values of $\rho(\hat{e}_i)$, we obtain $\text{STR}_{\text{reduced}} = 0.6111$. So $F_{\text{M}} = (0.6111 - 0.09676)/0.001485 = 346.5$.

The approximate p-value is $\text{Prob}[F \geq 346.5]$, where F is a random variable having an F distribution with 1 and 12 degrees of freedom. From the F table in the Appendix, we see that the p-value is less than 0.001. This verifies that the moisture content of the cereal increased significantly during storage.

5.5 AN EXAMPLE OF MULTIPLE REGRESSION

One measure of aerobic fitness is a person's rate of oxygen consumption. The faster the rate, the fitter the person. Table 5.3 lists the rate of oxygen consumption (in milliliters per kilogram of body weight per minute) for 31 individuals, together with measurements of five other variables: age (in years), weight (in kilograms), the time (in minutes) required to run $1\frac{1}{2}$ miles,

Table 5.3 Aerobic Fitness Data

Identification Number of the Individual	Oxygen Consumption (Y)	Age (X_1)	Weight (X_2)	Running Time (X_3)	Pulse at Rest (X_4)	Pulse After Run (X_5)
1	44.609	44	89.47	11.37	62	178
2	45.313	40	75.07	10.07	62	185
3	54.297	44	85.84	8.65	45	156
4	59.571	42	68.15	8.17	40	166
5	49.874	38	89.02	9.22	55	178
6	44.811	47	77.45	11.63	58	176
7	45.681	40	75.98	11.95	70	176
8	49.091	43	81.19	10.85	64	162
9	39.442	44	81.42	13.08	63	174
10	60.055	38	81.87	8.63	48	170
11	50.541	44	73.03	10.13	45	168
12	37.388	45	87.66	14.03	56	186
13	44.754	45	66.45	11.12	51	176
14	47.273	47	79.15	10.60	47	162
15	51.855	54	83.12	10.33	50	166
16	49.156	49	81.42	8.95	44	180
17	40.836	51	69.63	10.95	57	168
18	46.672	51	77.91	10.00	48	162
19	46.774	48	91.63	10.25	48	162
20	50.388	49	73.37	10.08	76	168
21	39.407	57	73.37	12.63	58	174
22	46.080	54	79.38	11.17	62	156
23	45.441	52	76.32	9.63	48	164
24	54.625	50	70.87	8.92	48	146
25	45.118	51	67.25	11.08	48	172
26	39.203	54	91.63	12.88	44	168
27	45.790	51	73.71	10.47	59	186
28	50.545	57	59.08	9.93	49	148
29	48.673	49	76.32	9.40	56	186
30	47.920	48	61.24	11.50	52	170
31	47.467	52	82.78	10.50	53	170

Source: SAS Institute Inc. (1990, p. 1443).

pulse rate (in beats per minute) while resting, and pulse rate at the end of the run. We can use these data to express rate of oxygen consumption as a function of the other variables. The linear regression model is $Y = \beta_0 + \beta_1 X_1 + \beta_2 X_2 + \beta_3 X_3 + \beta_4 X_4 + \beta_5 X_5 + e$.

First we make five plots of Y versus each of the explanatory variables X_j. None of these plots shows any definite indication that the regression function is nonlinear or that the errors have nonconstant variance.

5.6 ESTIMATING THE REGRESSION COEFFICIENTS

The procedure for finding M-estimates in multiple regression is a direct generalization of the procedure described in Section 5.3 for simple regression. The *Huber M-estimates* $\hat{\beta}_0, \hat{\beta}_1, \ldots, \hat{\beta}_p$ are the values of $b_0, b_1, \ldots, b_p$ that minimize

$$\sum \rho\left(y_i - (b_0 + b_1 x_{i1} + \cdots + b_p x_{ip})\right) \tag{5.4}$$

where $\rho(e)$ is the function defined in (5.1). It is convenient to use vector notation. Let

$$\boldsymbol{b} = \begin{bmatrix} b_0 \\ b_2 \\ \vdots \\ b_p \end{bmatrix} \quad \text{and} \quad \boldsymbol{x}_i = \begin{bmatrix} 1 \\ x_{i1} \\ \vdots \\ x_{ip} \end{bmatrix}$$

The vector $\hat{\boldsymbol{\beta}}$ of Huber M-estimates is defined to be the vector $\boldsymbol{b}$ that minimizes $\Sigma\rho(y_i - \boldsymbol{b}'\boldsymbol{x}_i)$.

An Algorithm. The vector of regression coefficients, denoted by $\boldsymbol{\beta}$, is first estimated by the vector of least-squares estimates. This initial estimate of $\boldsymbol{\beta}$ is used to calculate deviations and an estimate of σ. These in turn are used, as described in the next paragraph, to obtain an improved estimate of $\boldsymbol{\beta}$. The algorithm is iterated in this way until a step is reached at which the improved estimate of $\boldsymbol{\beta}$ is the same (or at least approximately the same) as the previous estimate.

To be specific, at any step in the algorithm, let $\boldsymbol{b}^0$ be the current estimate of $\boldsymbol{\beta}$. Calculate the deviations $y_i - (\boldsymbol{b}^0)'\boldsymbol{x}_i$ and from them calculate $\hat{\sigma} =$ 1.483MAD. Now make the following adjustment of the y-values to get rid of large deviations. The deviation of y_i from the current estimated regression line is $e_i^0 = y_i - (\boldsymbol{b}^0)'\boldsymbol{x}_i$. Thus $y_i = (\boldsymbol{b}^0)'\boldsymbol{x}_i + e_i^0$. Now define $y_i^* = (\boldsymbol{b}^0)'\boldsymbol{x}_i + e_i^*$, where e_i^* is the adjusted deviation obtained by truncating e_i^0 so that none of the deviations is larger than $1.5\hat{\sigma}$ in absolute value. Let the improved estimate of $\boldsymbol{\beta}$ be the least-squares estimate obtained from the adjusted data $y_1^*, \ldots, y_n^*$.

The Aerobic Fitness Data. For an initial estimate of the vector of regression coefficients we take the least-squares estimate, which is $\boldsymbol{b}^0 =$ (116.0, −0.2802, −0.05063, −2.743, −0.01224, −0.1279). Now we calculate the fitted y-values $\hat{y}_i^0 = (\boldsymbol{b}^0)'\boldsymbol{x}_i$ and the deviations $e_i^0 = y_i - \hat{y}_i^0$, as in

Table 5.4 Calculations in the First Iteration of the Algorithm for Obtaining M-Estimates for the Aerobic Fitness Data

Observed y-Value y_i	Fitted y-Value $\hat{y}_i^0$	Deviation e_i^0	Truncated Deviation e_i^*	Adjusted y-Value y_i^*
44.609	44.478	0.131	0.131	44.609
45.313	48.998	−3.685	−2.733	46.265
54.297	55.143	−0.846	−0.846	54.297
59.571	56.698	2.873	2.733	59.431
49.874	52.164	−2.290	−2.290	49.874
44.811	43.838	0.973	0.973	44.811
45.681	44.849	0.832	0.832	45.681
49.091	48.625	0.466	0.466	49.091
39.442	40.695	−1.253	−1.253	39.442
60.055	55.253	4.802	2.733	57.986
50.541	50.198	0.343	0.343	50.541
37.388	36.044	1.344	1.344	37.388
44.754	46.439	−1.685	−1.685	44.754
47.273	48.502	−1.229	−1.229	47.273
51.855	46.531	5.324	2.733	49.264
49.156	50.086	−0.930	−0.930	49.156
40.836	46.013	−5.177	−2.733	43.280
46.672	49.077	−2.405	−2.405	46.672
46.774	48.537	−1.763	−1.763	46.774
50.388	48.537	1.851	1.851	50.388
39.407	38.755	0.652	0.652	39.407
46.080	45.549	0.531	0.531	46.080
45.441	49.636	−4.195	−2.733	46.903
54.625	54.722	−0.097	−0.097	54.625
45.118	45.376	−0.258	−0.258	45.118
39.203	38.925	0.278	0.278	39.203
45.790	44.796	0.994	0.994	45.790
50.545	50.319	0.226	0.226	50.545
48.673	48.196	0.477	0.477	48.673
47.920	45.576	2.344	2.344	47.920
47.467	46.094	1.373	1.373	47.467

Table 5.4. The median of the absolute deviations $|e_i^0|$ is MAD $= 1.229$. So $\hat{\sigma} = (1.483)(1.229) = 1.822$ and $1.5\hat{\sigma} = 2.733$. Now truncate the deviations e_i^0 to obtain e_i^*. Three of the deviations are greater than $1.5\hat{\sigma}$, namely, $e_4^0 = 2.873$, $e_{10}^0 = 4.802$, and $e_{15}^0 = 5.324$. These are truncated to $e_4^* = e_{10}^* = e_{15}^* = 2.733$. Three deviations are less than $-1.5\hat{\sigma}$, namely, $e_2^0 = -3.685$, $e_{17}^0 = -5.177$, and $e_{23}^0 = -4.195$. These are truncated to $e_2^* = e_{17}^* = e_{23}^* = -2.733$. Now the adjusted y-values are obtained as $y_i^* = \hat{y}_i^0 + e_i^*$. Note that $y_i^* = y_i$ for all i except $i = 2, 4, 10, 15, 17$, and 23.

The vector of initial estimates was obtained by applying the least-squares method to the data in Table 5.3. If we replace the y-values in that table by the adjusted y-values from the last column of Table 5.4 and then apply the least-squares method of estimation, we obtain a new vector of estimates $\boldsymbol{b}^0 = (115.7, -0.2721, -0.07352, -2.694, -0.00059, -0.1245)$. This completes the first iteration of the algorithm.

For the second iteration we use the new vector of estimates $\boldsymbol{b}^0$ to calculate new fitted y-values $\hat{y}_i^0$. This constitutes column 2 in a new table similar to Table 5.4. After the last column of the new table is calculated, the least-squares method is applied to the new adjusted y-values to obtain an updated vector of estimates $\boldsymbol{b}^0 = (114.8, -0.2646, -0.07890, -2.678, 0.00601, -0.1228)$. Iteration is continued until the estimates converge. After 18 iterations, the relative difference between the updated estimate and the previous estimate is less than 10^{-4} for all six regression coefficients. Thus, to an accuracy of about four significant digits, we obtain $\hat{\boldsymbol{\beta}} = (113.1, -0.2489, -0.07718, -2.654, 0.01475, -0.1216)$. So the estimated regression equation is

$$\hat{Y} = 113.1 - 0.2489X_1 - 0.07718X_2 - 2.654X_3 + 0.01475X_4 - 0.1216X_5 \tag{5.5}$$

5.7 TESTING $\boldsymbol{\beta}_{q+1} = \cdots = \boldsymbol{\beta}_p = 0$

In the general linear regression model $Y = \beta_0 + \beta_1 X_1 + \cdots + \beta_p X_p + e$, recall that the least-squares test statistic for testing $\beta_{q+1} = \cdots = \beta_p = 0$ is

$$F_{\mathrm{LS}} = \frac{\mathrm{SSR}_{\mathrm{reduced}} - \mathrm{SSR}_{\mathrm{full}}}{(p-q)\hat{\sigma}_{\mathrm{LS}}^2}$$

where SSR stands for the sum of squared residuals, $\mathrm{SSR} = \Sigma \hat{e}_i^2$, and $\hat{\sigma}_{\mathrm{LS}}^2 = \Sigma \hat{e}_i^2/(n-p-1)$. The residuals in $\mathrm{SSR}_{\mathrm{reduced}}$ and $\mathrm{SSR}_{\mathrm{full}}$ are calculated by applying the least-squares method to, respectively, the reduced model $Y = \beta_0 + \beta_1 X_1 + \cdots + \beta_q X_q + e$ and the full model $Y = \beta_0 + \beta_1 X_1 + \cdots + \beta_p X_p + e$. In calculating $\hat{\sigma}_{\mathrm{LS}}^2$, the residuals from the full model are used.

A similar test statistic is used in M-regression:

$$F_{\mathrm{M}} = \frac{\mathrm{STR}_{\mathrm{reduced}} - \mathrm{STR}_{\mathrm{full}}}{(p-q)\hat{\lambda}} \tag{5.6}$$

where STR stands for the sum of transformed residuals, STR $= \Sigma\rho(\hat{e}_i)$, and $\hat{\lambda} = (n/m)\Sigma\hat{e}_i^{*2}/(n-p-1)$. The function $\rho(e)$ was introduced in formula (5.1) and $\hat{e}_i^*$ is obtained by truncating $\hat{e}_i$ as described in Section 5.3. That is, $\hat{e}_i^* = \hat{e}_i$ if $|\hat{e}_i| \le 1.5\hat{\sigma}$, $\hat{e}_i^* = -1.5\hat{\sigma}$ if $\hat{e}_i < -1.5\hat{\sigma}$, and $\hat{e}_i^* = 1.5\hat{\sigma}$ if $\hat{e}_i > 1.5\hat{\sigma}$. The estimate $\hat{\sigma}$ is 1.483MAD, as in the preceding section. The integer m is the number of residuals $\hat{e}_i$ that do not require truncation, that is, for which $|\hat{e}_i| \le 1.5\hat{\sigma}$. The residuals in $\text{STR}_{\text{reduced}}$ and STR_{full} are calculated by applying the M-regression procedure to, respectively, the reduced model and the full model. The estimation procedure for the reduced model differs slightly from the procedure described in the preceding section; the estimate of σ is not iterated. Throughout the iterations required to obtain the vector of M-estimates of the regression coefficients in the reduced model, we keep the same estimate $\hat{\sigma}$ calculated from the full model. In calculating $\hat{\lambda}$, the residuals from the full model are used.

An approximate p-value of the test is calculated in the same way as for the least-squares test, namely, as Prob$[F \ge F_{\text{M}}]$, where F denotes a random variable having an F distribution with $p-q$ and $n-p-1$ degrees of freedom.

Similarity to the Least-Squares Test. The formula for F_{M} is very similar to the formula for F_{LS}. In fact, if in the definition of $\hat{e}_i^*$ and $\rho(e)$, 1.5 is replaced by ∞, then F_{M} becomes exactly F_{LS}. Note that if 1.5 is replaced by ∞, then $\hat{e}_i^* = \hat{e}_i$ for all i and $m = n$, so that $\hat{\lambda}$ coincides with $\hat{\sigma}^2_{\text{LS}}$. Moreover, if 1.5 is replaced by ∞, then $\rho(e) = e^2$, so STR = SSR, and so F_{M} coincides with F_{LS}.

The Aerobic Fitness Data. Let us test whether the two measurements of pulse rate make a significant contribution in equation (5.5); that is, let us test the hypothesis that $\beta_4 = \beta_5 = 0$. We have $p = 5$ and $q = 3$.

To calculate F_{M} we first obtain the residuals $\hat{e}_i = y_i - \hat{\boldsymbol{\beta}}'\boldsymbol{x}_i$, where $\hat{\boldsymbol{\beta}}$ is the vector of M-estimates calculated in Section 5.6. The median of the absolute residuals $|\hat{e}_i|$ is MAD $= 0.9046$, so $\hat{\sigma} = (1.483)(0.9046) = 1.341$ and $1.5\hat{\sigma} = (1.5)(1.341) = 2.012$. For the 24 residuals between -2.012 and 2.012, $\rho(\hat{e}_i) = \hat{e}_i^2$, and for the other seven residuals, $\rho(\hat{e}_i) = 4.024|\hat{e}_i| - 4.049$. Adding all 31 values of $\rho(\hat{e}_i)$, we obtain $\text{STR}_{\text{full}} = 114.7$. By truncating the four residuals that are below -2.012 and the three residuals that are above 2.012, we obtain the truncated residuals $\hat{e}_i^*$. The sum of their squares is 50.44. Now $\hat{\lambda} = (31/24)(50.44)/25 = 2.606$.

The reduced model with $\beta_4 = \beta_5 = 0$ is $Y = \beta_0 + \beta_1 X_1 + \beta_2 X_2 + \beta_3 X_3 + e$. The Huber M-estimates are the values b_0, b_1, b_2, b_3 that minimize the function $\Sigma\rho(y_i - (b_0 + b_1 x_{i1} + b_2 x_{i2} + b_3 x_{i3}))$. The estimation procedure followed for the reduced model differs slightly from the procedure followed for the full model, because the estimate $\hat{\sigma}$ is not iterated. The same value $\hat{\sigma} = 1.341$, calculated from the full model, is used throughout the iterations required to obtain the M-estimates of the regression coefficients in the reduced model. So the value of k in the definition of ρ in (5.1) stays constant at $k = 1.5\hat{\sigma} = (1.5)(1.341) = 2.012$.

The vector of M-estimates in the reduced model is $(\hat{\beta}_0, \hat{\beta}_1, \hat{\beta}_2, \hat{\beta}_3) = (95.07, -0.1844, -0.08861, -3.015)$, so the residuals in the reduced model are $\hat{e}_i = y_i - (95.07 - 0.1844x_{i1} - 0.08861x_{i2} - 3.015x_{i3})$. For the 21 residuals between -2.012 and 2.012, $\rho(\hat{e}_i) = \hat{e}_i^2$, and for the other 10 residuals, $\rho(\hat{e}_i) = 4.024|\hat{e}_i| - 4.049$. Adding all 31 values of $\rho(\hat{e}_i)$, we obtain $\mathrm{STR}_{\mathrm{reduced}} = 145.8$. Now $F_{\mathrm{M}} = (145.8 - 114.7)/[(5 - 3)(2.606)] = 5.964$.

The approximate p-value of the test is Prob$[F \geq 5.964]$, where F is a random variable having an F distribution with 2 and 25 degrees of freedom. From the F table in the Appendix, we see that the p-value is between 0.001 and 0.01. We conclude that one or both of the pulse rate measurements have a significant relationship to oxygen consumption.

5.8 COMPUTATION

Huber M-estimates are obtainable from the function *rreg* in the statistical package S-PLUS (Becker, Chambers, and Wilks, 1988, p. 571) and from the robust regression packages ROBSYS (Marazzi, 1987) and ROSEPACK (Holland and Welsch, 1977).

If you have a computer package that performs least-squares regression, you can apply it iteratively to obtain Huber M-estimates. Let $\boldsymbol{y}$ denote the vector of observed response variables and $\boldsymbol{X}$ the matrix of explanatory variables. To begin, let $\boldsymbol{y}^{(1)} = \boldsymbol{y}$. Apply the package to $\boldsymbol{y}^{(1)}$ and $\boldsymbol{X}$ to obtain the vector $\boldsymbol{b}^{(1)}$ of least-squares regression estimates and the vector $\hat{\boldsymbol{y}}^{(1)}$ of predicted y-values. Let $\hat{\boldsymbol{e}}^{(1)} = \boldsymbol{y} - \hat{\boldsymbol{y}}^{(1)}$ be the vector of residuals. Calculate $\hat{\sigma}^{(1)} = 1.483\,\mathrm{median}(|\hat{e}_i^{(1)}|)$ and truncate the residuals by defining $f_i^{(1)} = \max(-1.5\hat{\sigma}^{(1)}, \min(\hat{e}_i^{(1)}, 1.5\hat{\sigma}^{(1)}))$. Let $\boldsymbol{y}^{(2)} = \hat{\boldsymbol{y}}^{(1)} + \boldsymbol{f}^{(1)}$ and repeat the calculations with $\boldsymbol{y}^{(2)}$ in place of $\boldsymbol{y}^{(1)}$. That is, apply the package to $\boldsymbol{y}^{(2)}$ and $\boldsymbol{X}$ to obtain $\boldsymbol{b}^{(2)}$ and $\hat{\boldsymbol{y}}^{(2)}$. Let $\hat{\boldsymbol{e}}^{(2)} = \boldsymbol{y} - \hat{\boldsymbol{y}}^{(2)}$ and then calculate $\hat{\sigma}^{(2)}$ and $\boldsymbol{f}^{(2)}$. Let $\boldsymbol{y}^{(3)} = \hat{\boldsymbol{y}}^{(2)} + \boldsymbol{f}^{(2)}$ and repeat the calculations with $\boldsymbol{y}^{(3)}$. Continue until consecutive estimates $\boldsymbol{b}^{(m-1)}$ and $\boldsymbol{b}^{(m)}$ are sufficiently close to one another; for example, continue until $|b_j^{(m-1)} - b_j^{(m)}| / |b_j^{(m-1)}| < 0.00001$ for all j. If the estimates $\boldsymbol{b}^{(1)}, \boldsymbol{b}^{(2)}, \ldots$ do not converge, try a different starting vector $\boldsymbol{b}^{(1)}$.

The test statistic F_{M} in (5.6) can be calculated as follows. Let $\hat{\boldsymbol{\beta}}$ be the vector of M-estimates, $\hat{\boldsymbol{y}} = \boldsymbol{X}\hat{\boldsymbol{\beta}}$, $\hat{\boldsymbol{e}} = \boldsymbol{y} - \hat{\boldsymbol{y}}$, $\hat{\sigma} = 1.483\,\mathrm{median}(|\hat{e}_i|)$, $f_i = \max(-1.5\hat{\sigma}, \min(\hat{e}_i, 1.5\hat{\sigma}))$, and $g_i = \max(0, |\hat{e}_i| - 1.5\hat{\sigma})$. Let $S_1 = \Sigma f_i^2$ and $S_2 = \Sigma g_i$. Then $\mathrm{STR}_{\mathrm{full}} = S_1 + 3\hat{\sigma}S_2$. Similar calculations using the reduced model produce $\text{STR}_{\text{reduced}}$. And $\hat{\lambda} = nS_1/(m(n - p - 1))$, where m is the number of g_i's equal to 0. Now apply formula (5.6).

Iteratively Reweighted Least Squares. The computational procedure used in ROBSYS, ROSEPACK, and S-PLUS is an iteratively reweighted least-squares procedure. This is a more general method than the one described in Section 5.6 and can be used to compute all types of M-estimates. For the case of Huber M-estimation, the procedure is described below.

To find the minimum of (5.4), for a fixed value of $\hat{\sigma}$, take its derivatives with respect to $b_0, b_1, \ldots, b_p$ and set them equal to 0. This gives us $p + 1$ equations in $p + 1$ unknowns:

$$\sum x_{ij}\rho'\big(y_i - (b_0 + b_1x_{i1} + \cdots + b_px_{ip})\big) = 0 \tag{5.7}$$

for $j = 0, 1, \ldots, p$, where we let $x_{i0} = 1$ for all i. These are nonlinear equations in the unknowns $b_0, b_1, \ldots, b_p$, but they can be approximated by linear equations as follows.

Consider an iterative procedure in which $b_0^0, b_1^0, \ldots, b_p^0$ are current estimates and $b_0, b_1, \ldots, b_p$ represent improved estimates. Let $e_i^0 = y_i - (b_0^0 + b_1^0x_{i1} + \cdots + b_p^0x_{ip})$ and $e_i = y_i - (b_0 + b_1x_{i1} + \cdots + b_px_{ip})$. To solve for the improved estimates, write $\rho'(e_i) = (\rho'(e_i)/e_i)e_i \approx (\rho'(e_i^0)/e_i^0)e_i$. Let $w_i = \rho'(e_i^0)/e_i^0$; that is,

$$w_i = \begin{cases} 2 & \text{if } |e_i^0| \le 1.5\hat{\sigma} \\ 3\hat{\sigma}/|e_i^0| & \text{if } |e_i^0| > 1.5\hat{\sigma} \end{cases}$$

Then $\rho'(e_i) \approx w_ie_i$ and we can approximate (5.7) by the linear equations

$$\sum x_{ij}w_i\big[y_i - (b_0 + b_1x_{i1} + \cdots + b_px_{ip})\big] = 0$$

Let $\boldsymbol{W}$ be the diagonal matrix with diagonal entries w_i. In terms of matrices we have $\boldsymbol{X}'\boldsymbol{W}(\boldsymbol{y} - \boldsymbol{X}\boldsymbol{b}) = 0$. Solving for $\boldsymbol{b}$, we obtain

$$\boldsymbol{b} = (\boldsymbol{X}'\boldsymbol{W}\boldsymbol{X})^{-1}\boldsymbol{X}'\boldsymbol{W}\boldsymbol{y} \tag{5.8}$$

To start, let $\boldsymbol{b}^0$ be the vector of least-squares estimates. At each iterative step, use the vector $\boldsymbol{b}^0$ of current estimates to calculate the vector $\boldsymbol{e}^0 = \boldsymbol{y} - \boldsymbol{X}\boldsymbol{b}^0$ of residuals. Then use the residuals to obtain $\hat{\sigma}$ and the weights w_i. The vector $\boldsymbol{b}$ of improved estimates can now be computed as in (5.8). Iterate until convergence.

The vector $\boldsymbol{b}$ in (5.8) is called a *weighted least-squares* estimate. Some least-squares regression packages also perform weighted regression. For example, the REG procedure in the SAS package has a WEIGHT statement. You could write a program in SAS to use the REG procedure iteratively to compute M-estimates. Or you could use the NLIN procedure in SAS, which has iteration built into it, with its _WEIGHT_ variable; see Example 5 for the NLIN procedure in the SAS/STAT User's Guide (1990).

Test Case. As a test case for M-regression computations, we can use the data in Table 3.2. For simple regression, let $X = X_1$. The estimated regression line is $\hat{Y} = 30.35 + 1.715X$. The estimate $\hat{\sigma}$ is 1.057. The test statistic for testing $\beta = 0$ is $|t_M| = 7.680$. For multiple regression, using both X_1 and X_2, the estimated regression equation is $\hat{Y} = 47.22 + 0.8285X_1 - 0.5080X_2$. The estimate $\hat{\sigma}$ is 2.133. The test statistic for testing $\beta_1 = \beta_2 = 0$ is $F_M = 7.923$.

NOTES

5.1a. The term "robust" is not precise, but a "robust" regression procedure generally refers to one that not only performs well if the population of errors is normally distributed but also is insensitive to small departures from the normality assumption. Huber (1964) constructed his M-estimate to be optimal if the error distribution is assumed to be a normal distribution contaminated by a small proportion of errors from some arbitrary distribution. His criterion for optimality was minimization of the maximum possible variance for infinitely large samples.

5.1b. The "M" in M-regression was chosen because of the relationship between M-estimation and maximum likelihood estimation. Maximum likelihood estimation is discussed in Chapter 10. Some (but not all) M-estimates would be maximum likelihood estimates if the population of errors were assumed to have a particular distribution. However, we do not make this assumption. In fact, the main point of M-estimation is to obtain estimates that perform well for a range of possible distributions for the population of errors.

5.3a. In (5.1), other values of k could be chosen. For a large value of k, $\rho(e) = e^2$ for most observed residuals e, and so the corresponding M-estimates are close to the LS estimates. For a small value of k, $\rho(e) = 2k|e| - k^2$ for most observed residuals e. Since minimizing $\Sigma(2k|\hat{e}_i| - k^2)$ is equivalent to minimizing $\Sigma|\hat{e}_i|$, the corresponding M-estimates are close to the LAD estimates. The value of k that gives the best estimates of the regression coefficients depends on the distribution of the population of random errors. If the distribution produces a large proportion of outliers, then a small value of k is best, whereas if the distribution produces a small proportion of outliers, such as the normal distribution does, then a large value of k is best. To guard against both possibilities, the value $k = 1.5\hat{\sigma}$ is a good choice.

Huber suggested $k = 1.5\hat{\sigma}$ in his 1981 book (p. 18). Holland and Welsch (1977, p. 818) suggested $k = 1.345\hat{\sigma}$. The latter value of k is chosen so that the M-estimates have an asymptotic efficiency (explained below) of 95% if the distribution of the errors is normal.

If it were known that the error distribution was normal, one would be well-advised to estimate α and β by least squares, because the LS estimates would be optimal in several senses (see Section 9.2). In particular, they would be asymptotically efficient, which means that, for very large samples, the LS estimates would be the most accurate. (In Section 9.2 it is mentioned that the LS estimates are maximum likelihood estimates when the error distribution is normal. Maximum likelihood estimation is asymptotically efficient for a wide variety of distributions.) But we do not know the true error distribution and so it may be prudent to use robust estimates, such as M-estimates, that can perform efficiently when the error distribution is not normal. Still, if we believe that the error distribution is close to normal, we do not want to sacrifice too much efficiency in case the error distribution is actually normal.

To say that a Huber M-estimate with $k = 1.345\hat{\sigma}$ has an asymptotic efficiency of 95% means that for an infinitely large sample, the reciprocal of the ratio of the variance of the Huber M-estimate to the smallest possible variance, which is the variance of the least-squares estimate when the error distribution is normal, is 0.95. As mentioned in Section 9.2, the asymptotic efficiency of the Huber M-estimate with $k = 1.5\hat{\sigma}$ is about 96%. In compensation for giving slightly higher efficiency in the case of a normal error distribution, $k = 1.5\hat{\sigma}$ gives slightly lower efficiency that $k = 1.345\hat{\sigma}$ in the case of an error distribution that is prone to outliers.

5.3b. In the definition of $\rho(e)$, $2k|e| - k^2$ is used instead of $|e|$ in order to make $\rho(e)$ a "smooth" function. It is smooth in the sense that it is a continuous function and has a continuous first derivative. The definition of $\rho(e)$ is in three parts: for e less than $-k$, for e between $-k$ and k, and for e greater than k. Let us check that $\rho(e)$ is continuous at the boundary points $e = -k$ and $e = k$. For e slightly less than $-k$, $\rho(e)$ is close to $2k|-k| - k^2 = k^2$; and for e slightly greater than $-k$, $\rho(e)$ is close to $(-k)^2 = k^2$; hence $\rho(e)$ is continuous at $e = -k$. Similarly, $\rho(e)$ can be shown to be continuous at $e = k$.

The derivative of $\rho(e)$ is, in the three parts of its definition, respectively, $\rho'(e) = -2k$, $\rho'(e) = 2e$, and $\rho'(e) = 2k$. Let us check that $\rho'(e)$ is continuous at the boundary points $e = -k$ and $e = k$. For e slightly less than $-k$, $\rho'(e)$ is equal to $-2k$; and for e slightly greater than $-k$, $\rho'(e)$ is close to $2(-k) = -2k$; hence $\rho'(e)$ is continuous at $e = -k$. Similarly, $\rho'(e)$ can be shown to be continuous at $e = k$.

5.3c. Other M-estimates can be obtained by using different functions ρ. The function ρ should qualify as a measure of "size". The general approach to estimation in linear regression analysis to choose estimates so that the residuals are "small" in some sense. The particular approach taken in M-estimation is to minimize $\Sigma\rho(\hat{e}_i)$, where $\rho(\hat{e}_i)$ is a measure of the "size" of the ith residual. A measure of size should satisfy the following properties: the residual 0 should have size 0; all sizes should be nonnegative; if one residual is farther from 0 than another residual, its size should be larger, or at least as large. That is, the function ρ should satisfy: $\rho(0) = 0$, $\rho(e) \geq 0$ for all e, $\rho(e_1) \geq \rho(e_2)$ whenever $e_1 < e_2 < 0$, and $\rho(e_1) \leq \rho(e_2)$ whenever $0 < e_1 < e_2$.

One could, for example, define $\rho(e) = \log(1 + e^2)$. The M-estimate for this choice of ρ would be a maximum likelihood estimate if the error population had a Cauchy distribution. (See Note 5.1b.) Another example is Tukey's biweight M-estimate using $\rho(e) = (3k^4e^2 - 3k^2e^4 + e^6)/6$ for $-k \leq e \leq k$ and $= k^6/6$ for $e \leq -k$ and $e \geq k$. (It is usually presented in terms of the derivative $\psi(e) = \rho'(e) = e(k^2 - e^2)^2$ for $-k \leq e \leq k$ and $= 0$ for $e \leq -k$ and $e \geq k$. See Note 5.3e.) This M-estimate is not the maximum likelihood estimate for any distribution of the error population; see Note 5.3d.

5.3d. An M-estimate obtained using a function ρ would be the maximum likelihood estimate if the distribution of the error population had the probability density function $f(t) = ce^{-\rho(t)}$, where e denotes the base of the natural logarithms, $e \approx 2.178$ (so we have changed the variable e in $\rho(e)$ to t). The sum $\Sigma\rho(e_i)$ is minimized if and only if the likelihood function $\Pi f(e_i) = c^n e^{-\Sigma\rho(e_i)}$ is maximized. In order for $f(t)$ to be a valid p.d.f., c must equal the reciprocal of the integral $\int_{-\infty}^{\infty} e^{-\rho(t)}\,dt$, so that the area under the curve $f(t)$ is 1, and the integral must be finite. The integral is finite for the Huber ρ function and for the function $\rho(t) = \log(1 + t^2)$ in Note 5.3c. The integral is infinite for Tukey's biweight ρ function, which is a constant A $(= k^6/6)$ for $e \geq k$, because $\int_{-\infty}^{\infty} e^{-\rho(t)}\,dt \geq \int_k^{\infty} e^{-A}\,dt = \infty$.

5.3e. M-estimates can be presented in terms of the derivative $\psi(e) = \rho'(e)$ rather than $\rho(e)$. The function ψ is often preferred, because it determines the shape of the influence function of the estimate; see Notes 5.6e and 5.6f. Instead of minimizing (5.2) involving ρ, one can solve equations (5.3) involving ψ. The ψ function for the Huber M-estimate, after division by 2, is $\psi(e) = -k$ if $e < -k$, $= e$ if $-k \leq e \leq k$, and $= k$ if $e > k$. Note that an M-estimate remains the same if ψ is divided by a constant, because the solution of (5.3) remains the same.

5.3f. We use $\hat{\sigma} = 1.483\text{MAD}$ to estimate the standard deviation σ of the distribution of the random errors. If the distribution is normal, then this estimate has the following property. When the formula $\hat{\sigma} = 1.483\text{MAD}$ is applied to very large samples, $\hat{\sigma}$ tends to be very close to σ; that is, $\hat{\sigma}$ is a *consistent* estimator of σ. However, for a sample of size $n = 14$, which is not large, we cannot necessarily expect $\hat{\sigma}$ to be very close to σ, even if the distribution of errors were normal.

5.3g. Another common method for estimating σ in the context of M-estimation is Huber's Proposal 2. See Huber (1981, Section 7.7) and Rocke and Shanno (1986). To describe this method let us first express the Huber M-estimates of the regression coefficients in terms of standardized deviations. It is convenient also to express the M-estimates in terms of the function $\psi = \rho'$ rather than the function ρ (see Note 5.3e). The Huber M-estimates of α and β are the values of a and b that solve the equations $\Sigma\psi(e_i) = 0$ and $\Sigma x_i\psi(e_i) = 0$ (see (5.3)), where e_i denotes the deviation $y_i - (a + bx_i)$ and $\psi(e) = -1.5\hat{\sigma}$ if $e < -1.5\hat{\sigma}$, $= e$ if $-1.5\hat{\sigma} \le e \le 1.5\hat{\sigma}$, and $= 1.5\hat{\sigma}$ if $e > 1.5\hat{\sigma}$. These equations can be put in terms of standardized deviations: (1) $\Sigma\psi_0(z_i) = 0$ and (2) $\Sigma x_i\psi_0(z_i) = 0$, where z_i denotes the standardized deviation $e_i/\hat{\sigma}$ and $\psi_0(z) = -1.5$ if $z < -1.5$, $= z$ if $-1.5 \le z \le 1.5$, and $= 1.5$ if $z > 1.5$.

In Section 5.3 we solve (1) and (2) for a and b with $\hat{\sigma}$ fixed. In Huber's Proposal 2, we replace $\hat{\sigma}$ by the variable s in equations (1) and (2), letting $z_i = e_i/s$, add a third equation (3) $\Sigma(\psi_0(z_i))^2 = 0.7785n$, and solve equations (1), (2) and (3) simultaneously for a, b and s. The solutions are the M-estimates $\hat{\alpha}$, $\hat{\beta}$ and $\hat{\sigma}$.

To make some sense out of equation (3), consider what happens if 1.5 is replaced by ∞. Then $\psi_0(z) = z$ and the solutions of equations (1) and (2) are the least-squares estimates of α and β (see Note 3.3a). Equation (3) becomes $s^2 = \Sigma((y_i - (a + bx_i))^2/(0.7785n)$; putting $a = \hat{\alpha}_{\text{LS}}$ and $b = \hat{\beta}_{\text{LS}}$, we get $s^2 = \hat{\sigma}^2_{\text{LS}}((n-2)/n)/0.7785 \approx \hat{\sigma}^2_{\text{LS}}/0.7785$. Apart from the number 0.7785, this shows how equation (3) is related to estimation of σ. Now it remains to explain why 0.7785 appears in equation (3) for the Huber ψ_0 function with 1.5. Consider the standardized residuals $z_i = (y_i - (\alpha + \beta x_i))/\sigma$ using the true parameter values α, β and σ. For large sample sizes, $(1/n)\Sigma(\psi_0(z_i))^2$ should be close to the expectation $E[(\psi_0(z))^2]$, where $z = e/\sigma$ and e is randomly chosen from the error population. (This is a consequence of the Law of Large Numbers, which says that if a large sample is randomly selected from a population, then the sample average should be close to the population average.) In the case of a normally distributed error population, z has the standard normal distribution and one can calculate that $E[(\psi_0(z))^2] = 0.7785$. Another way of saying this is that 0.7785 makes the M-estimate $\hat{\sigma}$ a consistent estimate of σ when the error distribution is normal. (A "consistent" estimate is one that, for very large samples, tends to be very close to the parameter being estimated.)

5.3h. As the initial estimates of α and β in the algorithm it is natural to use the least-squares estimates, because at each iteration the algorithm uses the method of least squares to calculate the improved estimates. However, convergence of the algorithm may be facilitated by using initial estimates that are more robust, such as LAD estimates or least-median-of-squares estimates (see Rousseeuw and Leroy, 1987).

5.3i. In the algorithm presented here, the estimate $\hat{\sigma}$ is updated at every iteration. This is the procedure followed in the computer packages ROBSYS (Marazzi, 1987) and S-PLUS (Becker, Chambers, and Wilks, 1988). But ROSEPACK (Holland and Welsch, 1977) uses the LAD estimates as initial estimates of the regression coefficients, calculates $\hat{\sigma} = 1.483\text{MAD}$, and then keeps this same value of $\hat{\sigma}$ throughout the iterations that are required to converge to the M-estimates of the regression coefficients. By not updating the estimate of σ, the amount of calculation is reduced. This is a definite advantage in simulation studies in which the estimation procedure must be repeated thousands of times. But if the initial estimates of the regression coefficients is poor, then an estimate of σ based on them may well be poor, and this may adversely affect the robustness of the procedure (Shanno and Rocke, 1986, p. 88).

5.3j. A convenient criterion to determine when to stop iterating is the following. Stop when two successive pairs of estimates, say, (a^0, b^0) and (a^1, b^1), satisfy the condition that both relative differences $|a^1 - a^0|/\,|a^0|$ and $|b^1 - b^0|/\,|b^0|$ are less than 10^{-4}. This guarantees that the two successive estimates "almost" agree to four significant digits (or to k significant digits if 10^{-4} is replaced by 10^{-k}). To explain what is meant by "almost", round a^0 to four significant digits and regard these four digits as an integer m^0. The integer m^0 is between 1000 and 10,000. Similarly, round a^1 to the same number of decimal places as a^0 (which implies that a^1 is also rounded to four significant digits if a^1 is close to a^0) and regard its digits as an integer m^1. when we say that a^0 and a^1 agree to four significant digits, we mean that $m^0 = m^1$. When we say that a^0 and a^1 "almost" agree to four significant digits, we mean that either m^0 and m^1 are equal or they differ by 1.

However, even if two successive estimates agree to k significant digits, this does not guarantee that the M-estimate is accurate to k significant digits. See Note 5.3k.

5.3k. Even though two successive estimates agree to four significant digits, this may not yield an M-estimate that is accurate to four significant digits. This problem can occur when the function (5.2), which we are trying to minimize, is rather "flat". To increase our confidence that our M-estimates are accurate to four significant digits, there are two precautions that can be taken. (1) We can iterate until both relative differences $|a^1 - a^0|/\,|a^0|$ and

$|b^1 - b^0| / |b^0|$ are less than 10^{-k} with $k = 5$ or more. For the shelf life data, we tried $k = 8$ and, after 16 iterations, obtained the same results, when rounded to four significant digits, as with $k = 4$. (2) We can try several different initial estimates and see if the algorithm converges to the same four significant digits.

5.4. The convergence of the M-estimation procedure for the shelf life data using the model $Y = \alpha + e$ is very slow when we use the least-squares estimate $a^0 = \bar{y}$ as the initial estimate. To remedy slow convergence, a different initial estimate can be tried. We tried the LAD estimate. For this model, the LAD estimate of α is simply the sample median, 3.45. The M-estimation procedure converges immediately when 3.45 is the initial estimate because all values of a between $3.4 + 0.04574 = 3.446$ and $3.5 - 0.04574 = 3.454$ minimize the function $\Sigma\rho(y_i - a)$, where ρ is defined as in (5.1) with $k = 0.04574$.

5.6a. An M-estimate can be defined in terms of the function ρ as in (5.4) or in terms of the derivative $\psi = \rho'$. The function ψ is often preferred, because it determines the shape of the influence function of the estimate; see Notes 5.6e and 5.6f. One can take $\hat{\boldsymbol{\beta}}$ to be the value of $\boldsymbol{b}$ that minimizes $\Sigma\rho(y_i - \boldsymbol{b}'\boldsymbol{x}_i)$ or, equivalently, the value of $\boldsymbol{b}$ for which the partial derivatives $(\partial/\partial b_j)[\rho(y_i - \boldsymbol{b}'\boldsymbol{x}_i)]$ are zero for all $j = 0, 1, \ldots, p$. The vector of partial derivatives is $-\Sigma\psi(y_i - \boldsymbol{b}'\boldsymbol{x}_i)\boldsymbol{x}_i$, and so $\hat{\boldsymbol{\beta}}$ can be defined by the equation $\Sigma\psi(y_i - \hat{\boldsymbol{\beta}}'\boldsymbol{x}_i)\boldsymbol{x}_i = \mathbf{0}$.

Consider the case when $\rho(e) = e^2$, that is, when $\hat{\boldsymbol{\beta}} = \hat{\boldsymbol{\beta}}_{LS}$. Then $\psi(e) = 2e$, and so, after dividing by 2, the equation becomes $\Sigma(y_i - \hat{\boldsymbol{\beta}}'_{LS}\boldsymbol{x}_i)\boldsymbol{x}_i = \mathbf{0}$. It can be rewritten as $\Sigma\boldsymbol{x}_i(y_i - \boldsymbol{x}'_i\hat{\boldsymbol{\beta}}_{LS}) = \mathbf{0}$ and then as $\boldsymbol{X}'\boldsymbol{y} - \boldsymbol{X}'\boldsymbol{X}\hat{\boldsymbol{\beta}}_{LS} = \mathbf{0}$, which agrees with (3.8).

5.6b. As the initial estimates of the regression coefficients in the algorithm we have used the least-squares estimates, but a better choice might be the LAD estimates. See Note 5.3h.

5.6c. Other methods of estimating σ have also been suggested. Huber's Proposal 2, described in Note 5.3g for simple regression, can also be applied in multiple regression; simply replace $a + bx_i$ by $\boldsymbol{b}'\boldsymbol{x}_i$. See also Notes 5.3i and 5.7e.

5.6d. We iterated the estimation procedure until two successive vectors of estimates, say, $(b_0^0, b_1^0, b_2^0, b_3^0, b_4^0, b_5^0)$ and $(b_0^1, b_1^1, b_2^1, b_3^1, b_4^1, b_5^1)$, satisfied the condition that the relative differences $|b_j^1 - b_j^0| / |b_j^0|$ were less than 10^{-4} for all six regression coefficients. This guarantees that the two successive estimates "almost" agree to four significant digits. See Note 5.3j. However, such agreement does not guarantee that the M-estimates have this degree of accuracy. See Note 5.3k. To increase our confidence in the accuracy of our M-estimate for the aerobic fitness data, we continued the iterations until the relative differences between successive estimates were less than 10^{-8} for all six regression coefficients. After 36 iterations, we obtained the same results, when rounded to four significant digits, as with 10^{-4}.

5.6e. The concept of the influence function of an estimate is central in robust statistics. To define this concept, we first consider the simplest kind of data, with no explanatory variables. Suppose $y_1, \ldots, y_n$ are a sample of numbers that have been randomly selected from a population whose distribution depends on an unknown parameter θ. Let $T(y_1, \ldots, y_n)$ be an estimate of θ calculated from the sample. How much would the estimate change if we randomly selected one more number from the population? The difference $T(y_1, \ldots, y_n, z) - T(y_1, \ldots, y_n)$ provides a measure of the influence that the additional number would have on the estimate if the value of the number were z. But to pursue a mathematical analysis of influence, it is convenient to modify this measure as follows.

First we express the estimate $T(y_1, \ldots, y_n)$ as a function $T(P_n)$ of the empirical probability distribution P_n that assigns probability $1/n$ to each y_i in the sample. For example, consider the sample mean $\bar{y} = (1/n)\Sigma y_i$. Note that $\bar{y}$ is the mean of the distribution P_n, and so $\bar{y} = T(P_n)$ where, for any probability distribution P on the real line, $T(P)$ is defined to be its mean (provided it exists).

Let P_{n+1}^z denote the empirical probability distribution for the sample with z appended; that is, $P_{n+1}^z(y_i) = 1/(n+1)$ and $P_{n+1}^z(z) = 1/(n+1)$. Let δ^z denote the "degenerate" probability function that assigns all its probability to z; that is, $\delta^z(z) = 1$. Note that $P_{n+1}^z = (n/(n+1))P_n + (1/(n+1))\delta^z$. The influence of z on $T(P_n)$ could be expressed as $T((1-\epsilon_n)P_n + \epsilon_n\delta^z) - T(P_n)$, where $\epsilon_n = 1/(n+1)$. Modifying this somewhat, we define the *influence function* of the estimate $T(y_1, \ldots, y_n)$ to be $\mathrm{IF}(z) = \lim_{\epsilon \to 0}\{T((1-\epsilon)P_n + \epsilon\delta^z) - T(P_n)\}/\epsilon$, which is the derivative of $T((1-\epsilon)P_n + \epsilon\delta^z)$ with respect to ϵ at $\epsilon = 0$. Thus the influence function is the rate of change of the estimate when a small proportion of additional data with value z is included in the sample.

For the sample mean, $T((1-\epsilon)P_n + \epsilon\delta^z) = (1-\epsilon)T(P_n) + \epsilon T(\delta^z) = (1-\epsilon)\bar{y} + \epsilon z = \bar{y} + \epsilon(z - \bar{y})$, and so $\mathrm{IF}(z) = z - \bar{y}$. This indicates that the sample mean is not robust. For an estimate to be regarded as robust against outliers, its influence function must be bounded; a small proportion of additional data should not be allowed to cause arbitrarily large changes in the estimate.

Now consider a sample of regression data $(\boldsymbol{x}_1, y_1), \ldots, (\boldsymbol{x}_n, y_n)$. We assume the linear regression model $y_i = \boldsymbol{\beta}'\boldsymbol{x}_i + e_i$. (It is convenient to include 1 as the first component of $\boldsymbol{x}_i$ so that $\boldsymbol{x}_i = (1, x_{i1}, \ldots, x_{ip})$ and $\boldsymbol{\beta}'\boldsymbol{x}_i = \beta_0 + \beta_1 x_{i1} + \cdots + \beta_p x_{ip}$.) Let $\boldsymbol{T}(\boldsymbol{x}_1, y_1, \ldots, \boldsymbol{x}_n, y_n)$ be an estimate of $\boldsymbol{\beta}$ calculated from the sample. To define the influence function of the estimate, we regard the explanatory variables, as well as the response variables, as being random. Let P_n denote the empirical probability distribution that assigns probability $1/n$ to each data point $(\boldsymbol{x}_i, y_i)$, and express $\boldsymbol{T}(\boldsymbol{x}_1, y_1, \ldots, \boldsymbol{x}_n, y_n)$ as a function of P_n. For example, the least-squares estimate of $\boldsymbol{\beta}$ can be written as $\hat{\boldsymbol{\beta}}_{\text{LS}} = \boldsymbol{T}(P_n)$ where $\boldsymbol{T}(P) = [E_P(\boldsymbol{x}\boldsymbol{x}')]^{-1}E_P(\boldsymbol{x}y)$, in which $(\boldsymbol{x}, y)$ is a random vector with distribution P.

The *influence function* $\text{IF}(\boldsymbol{\omega}, z)$ of the estimate $\boldsymbol{T}(\boldsymbol{x}_1, y_1, \ldots, \boldsymbol{x}_n, y_n)$ is defined to be the vector of derivatives of $\boldsymbol{T}((1 - \epsilon)P_n + \epsilon\delta^{(\boldsymbol{w}, z)})$ with respect to ϵ at $\epsilon = 0$. Thus $\text{IF}(\boldsymbol{w}, z)$ gives the rate of change of the estimate when a small proportion of additional data with values $(\boldsymbol{w}, z)$ is included in the sample. For the least-squares estimate, $\text{IF}(\boldsymbol{w}, z) = n(\boldsymbol{X}'\boldsymbol{X})^{-1}\boldsymbol{w}(z - \hat{\boldsymbol{\beta}}'_{\text{LS}}\boldsymbol{w})$ (see Staudte and Sheather, 1990, Section 7.4.3).

5.6f. The influence function of an M-estimate of $\boldsymbol{\beta}$ is $\text{IF}(\boldsymbol{w}, z) = \boldsymbol{M}^{-1}\boldsymbol{w}\psi(z - \hat{\boldsymbol{\beta}}'\boldsymbol{w})$, where $\boldsymbol{M} = (1/n)\Sigma\psi'(y_i - \hat{\boldsymbol{\beta}}'\boldsymbol{x}_i)\boldsymbol{x}_i\boldsymbol{x}_i'$ and $\psi = \rho'$. The influence function for each component of the M-estimate, as a function of z for fixed $\boldsymbol{w}$, has the form $\text{IF}(z) = a\psi(z - c)$ for constants a and c. Thus the shape of the influence function of an M-estimate is determined by its ψ function.

The influence function is derived as follows. As seen in Note 5.6a, the M-estimate $\hat{\boldsymbol{\beta}}$ is defined implicitly by the equation $\Sigma\psi(y_i - \hat{\boldsymbol{\beta}}'\boldsymbol{x}_i)\boldsymbol{x}_i = \boldsymbol{0}$. In order to apply the definition of influence function presented in Note 5.6e, we express the estimate as $\hat{\boldsymbol{\beta}} = \boldsymbol{T}(P_n)$, where $\boldsymbol{T}(P)$ is defined implicitly by the equation $E_P[\psi(y - \boldsymbol{T}(P)'\boldsymbol{x})\boldsymbol{x}] = \boldsymbol{0}$, in which $(\boldsymbol{x}, y)$ is a random vector with distribution P. For an arbitrary vector $(\boldsymbol{w}, z)$, let $P_\epsilon = (1 - \epsilon)P_n + \epsilon\delta^{(\boldsymbol{w}, z)}$. The influence function $\text{IF}(\boldsymbol{w}, z)$ is defined to be the vector of derivatives of $\boldsymbol{T}(P_\epsilon)$ with respect to ϵ at $\epsilon = 0$.

First note that for any function h, $E_{P_\epsilon}[h(\boldsymbol{x}, y)] = (1 - \epsilon)E_{P_n}[h(\boldsymbol{x}, y)] + \epsilon E_\delta[h(\boldsymbol{x}, y)] = (1 - \epsilon)(1/n)\Sigma h(\boldsymbol{x}_i, y_i) + \epsilon h(\boldsymbol{w}, z)$. Hence $\boldsymbol{0} = E_{P_\epsilon}[\psi(y - \boldsymbol{T}(P_\epsilon)'\boldsymbol{x})\boldsymbol{x}] = (1 - \epsilon)(1/n)\Sigma\psi(y_i - \boldsymbol{T}(P_\epsilon)'\boldsymbol{x}_i)\boldsymbol{x}_i + \epsilon\psi(z - \boldsymbol{T}(P_\epsilon)'\boldsymbol{w})\boldsymbol{w}$. Differentiate with respect to ϵ, set $\epsilon = 0$, and note that $P_\epsilon = P_n$ and $\boldsymbol{T}(P_\epsilon) = \hat{\boldsymbol{\beta}}$ when $\epsilon = 0$. This leads to $\boldsymbol{0} = -(1/n)\Sigma\psi'(y_i - \hat{\boldsymbol{\beta}}'\boldsymbol{x}_i)(\text{IF}(\boldsymbol{w}, z)'\boldsymbol{x}_i)\boldsymbol{x}_i + \psi(z - \hat{\boldsymbol{\beta}}'\boldsymbol{w})\boldsymbol{w}$, which can be solved for $\text{IF}(\boldsymbol{w}, z)$. For the Huber ρ function in (5.1), $\psi'(e) = 2$ for $|e| < k$ and $= 0$ for $|e| > k$.

5.7a. Test statistic (5.6) was introduced by Schrader and Hettmansperger (1980).

5.7b. Test statistic (5.6) should be close to 1 if the null hypothesis is true. To see why, let us approximate the expectation of the numerator, $\text{STR}_{\text{reduced}} - \text{STR}_{\text{full}}$, assuming that the null hypothesis is true. Consider $\text{STR}_{\text{full}} = \Sigma\rho(y_i - \hat{\boldsymbol{\beta}}'\boldsymbol{x}_i)$, where $\hat{\boldsymbol{\beta}}$ is the M-estimate determined by ρ. (In this note, ρ could be any function described in Note 5.3c—not necessarily Huber's ρ function.) As a function of $\boldsymbol{\beta}$, $\rho(y - \boldsymbol{\beta}'\boldsymbol{x})$ can be approximated by a quadratic Taylor polynomial, $\rho(y - \boldsymbol{\beta}'\boldsymbol{x}) \approx \rho(y - \boldsymbol{\beta}_0'\boldsymbol{x}) - \psi(y - \boldsymbol{\beta}_0'\boldsymbol{x})\boldsymbol{x}'(\boldsymbol{\beta} - \boldsymbol{\beta}_0)$

$+ \frac{1}{2}\psi'(y - \boldsymbol{\beta}_0'\boldsymbol{x})(\boldsymbol{\beta} - \boldsymbol{\beta}_0)'\boldsymbol{x}\boldsymbol{x}'(\boldsymbol{\beta} - \boldsymbol{\beta}_0)$, where $\psi = \rho'$ as in Note 5.6a. Apply this to $\rho(y_i - \boldsymbol{\beta}'\boldsymbol{x}_i)$ and sum over i. Let $\boldsymbol{\beta}$ be the true parameter vector, let $\boldsymbol{\beta}_0 = \hat{\boldsymbol{\beta}}$, and use the fact that $\Sigma\psi(y_i - \hat{\boldsymbol{\beta}}'\boldsymbol{x}_i)\boldsymbol{x}_i = 0$ (see Note 5.6a). This yields the approximation $\text{STR}_{\text{full}} \approx \Sigma\rho(e_i) - \frac{1}{2}(\hat{\boldsymbol{\beta}} - \boldsymbol{\beta})'[\Sigma\psi'(\hat{e}_i)\boldsymbol{x}_i\boldsymbol{x}_i'](\hat{\boldsymbol{\beta}} - \boldsymbol{\beta})$, where $e_i = y_i - \boldsymbol{\beta}'\boldsymbol{x}_i$ and $\hat{e}_i = y_i - \hat{\boldsymbol{\beta}}'\boldsymbol{x}_i$.

If we suppose that the residuals $\hat{e}_i$ are approximately independent of the estimate $\hat{\boldsymbol{\beta}}$ (see Sheather and McKean, 1992, Table 1 and Figure 1), then $E(\text{STR}_{\text{full}}) \approx nE[\rho(e)] - \frac{1}{2}E[\psi'(e)]\,\text{trace}[\boldsymbol{X}'\boldsymbol{X}\text{Cov}(\hat{\boldsymbol{\beta}})]$, where e is randomly drawn from the population of errors. According to Note 5.7d, $\text{Cov}(\hat{\boldsymbol{\beta}})] \approx \tau^2(\boldsymbol{X}'\boldsymbol{X})^{-1}$, where $\tau^2 = E\left[\psi^2(e)\right]\big/\left(E\left[\psi'(e)\right]\right)^2$. (The parameter τ^2 plays a role in M-regression similar to the role of σ^2 in least-squares regression. Check that $\tau^2 = \sigma^2$ when $\rho(e) = e^2$.) Therefore $E(\text{STR}_{\text{full}}) \approx nE[\rho(e)] - \lambda(p + 1)$, where $\lambda = \frac{1}{2}E\left[\psi^2(e)\right]\big/E\left[\psi'(e)\right]$. If the null hypothesis is true, then $E(\text{STR}_{\text{reduced}}) \approx nE[\rho(e)] - \lambda(q + 1)$, and so $E(\text{STR}_{\text{reduced}} - \text{STR}_{\text{full}}) \approx (p - q)\lambda$. The denominator of (5.6) is an estimate of $(p - q)\lambda$ (see Note 5.7c). Therefore, if the null hypothesis is true, the ratio of numerator to denominator should be close to 1.

5.7c. The quantity $\hat{\lambda} = (n/m)\Sigma\hat{e}_i^{*2}/(n - p - 1)$ in (5.6) is an estimate of the parameter $\lambda = \frac{1}{2}E[\psi^2(e)]/E[\psi'(e)]$ that occurs in Note 5.7b. A simple estimate of $E[\psi^2(e)]$ is $(1/n)\Sigma\psi^2(\hat{e}_i)$ and a simple estimate of $E[\psi'(e)]$ is $(1/n)\Sigma\psi'(\hat{e}_i)$. For the Huber ψ function, $\psi(\hat{e}_i) = 2\hat{e}_i^*$ and $\psi'(\hat{e}_i) = 2$ if $|\hat{e}_i| < 1.5\hat{\sigma}, = 0$ if $|\hat{e}_i| > 1.5\hat{\sigma}$. Therefore a reasonable estimate of λ is $\Sigma\hat{e}_i^{*2}/m$, where m is the number of residuals $\hat{e}_i$ such that $|\hat{e}_i| < 1.5\hat{\sigma}$. It has been found that the accuracy of the p-value of the test is improved by multiplying the estimate by $n/(n - p - 1)$.

5.7d. The distribution of the M-estimate $\hat{\boldsymbol{\beta}}$ cannot be specified exactly, but for large sample sizes, under certain assumptions, the distribution is approximately normal with mean vector $\boldsymbol{\beta}$ and variance-covariance matrix $\tau^2(\boldsymbol{X}'\boldsymbol{X})^{-1}$ where $\tau^2 = E[\psi^2(e)]/(E[\psi'(e)])^2$.

More generally, a wide variety of esimates have distributions that are approximately normal with mean vector $\boldsymbol{\beta}$ and a variance-covariance matrix that can be described in terms of the influence function. In Note 5.6e the influence function of an estimate $\hat{\boldsymbol{\beta}} = \boldsymbol{T}(P_n)$ is defined to be the vector of derivatives of $\boldsymbol{T}((1 - \epsilon)P_n + \epsilon\delta^{(\boldsymbol{x}, y)})$ with respect to ϵ at $\epsilon = 0$. Denote it by $\text{IF}(\boldsymbol{x}, y; \boldsymbol{T}, P_n)$. In the same way, define $\text{IF}(\boldsymbol{x}, y; \boldsymbol{T}, P)$ by replacing P_n by the true (unknown) probability distribution P of $(\boldsymbol{x}, y)$. If certain assumptions are met, then $\text{Cov}(\hat{\boldsymbol{\beta}}) \approx (1/n)E[\text{IF}(\boldsymbol{x}, y; \boldsymbol{T}, P) \cdot \text{IF}(\boldsymbol{x}, y; \boldsymbol{T}, P)']$. (See Huber, 1981, p. 14 and Sections 7.4 and 7.6; Maronna and Yohai, 1981; Hampel et al., 1986, Section 6.3a.)

In Note 5.6f it is shown that the influence function of an M-estimate is $\text{IF}(\boldsymbol{x}, y; \boldsymbol{T}, P_n) = \boldsymbol{M}_n^{-1}\boldsymbol{x}\psi(y - \boldsymbol{T}(P_n)'\boldsymbol{x})$, where $\boldsymbol{M}_n = (1/n)\Sigma\psi'(y_i - \hat{\boldsymbol{\beta}}'\boldsymbol{x}_i)\boldsymbol{x}_i\boldsymbol{x}_i'$. Similarly, $\text{IF}(\boldsymbol{x}, y; \boldsymbol{T}, P) = \boldsymbol{M}^{-1}\boldsymbol{x}\psi(y - \boldsymbol{T}(P)'\boldsymbol{x}) = \boldsymbol{M}^{-1}\boldsymbol{x}\psi(e)$, where $M = E[\psi'(e)xx'] = E[\psi'(e)]E[xx']$. Therefore $\text{Cov}(\hat{\boldsymbol{\beta}}) \approx (1/n)M^{-1}E[\psi^2(e)xx']M^{-1} = (\tau^2/n)(E[xx'])^{-1}$. A natural estimate of $E[xx']$ is $(1/n)\Sigma x_i x_i' = (1/n)X'X$. Hence $\text{Cov}(\hat{\boldsymbol{\beta}}) \approx \tau^2(X'X)^{-1}$.

5.7e. To estimate σ for the purpose of testing a hypothesis about the regression coefficients, as distinct from the purpose of estimating the regression coefficients, other methods have also been used. When estimating, the main concern is that the estimate be close to the parameter being estimated. When testing, the main concerns are that the p-values are accurate and that the test has good power (that is, the p-values tend to be large when the null hypothesis if false). For estimation purposes, the two most common estimates of σ are 1.483MAD, which is the estimate used in this chapter, and Huber's Proposal 2 (see Note 5.6c). For testing purposes, Schrader and Hettmansperger (1980, Section 3.2) found that Huber's Proposal 2 works well but that $\hat{\sigma} = 1.483\text{MAD}$ yields tests with inaccurate p-values. They proposed $\hat{\sigma}^* = 2.1\text{MAD}^*$, where MAD* is the median of the $n - p$ largest deviations based on the LAD estimates of the regression coefficients.

We should explain why, in spite of the findings of Schrader and Hettmansperger, we have used $\hat{\sigma} = 1.483\text{MAD}$. In their paper they used Huber M-estimates with $k = 1.0\hat{\sigma}^* = 2.1\text{MAD}^*$ and $k = 1.2\hat{\sigma}^* = (1.2)(2.1\text{MAD}^*) = 2.52\text{MAD}^*$. We have used $k = 1.5\hat{\sigma} = (1.5)(1.483\text{MAD}) = 2.225\text{MAD}$. So our procedure is roughly the same as theirs, and ours is more "natural" in that the same estimator is used in estimation and testing and that M-estimates rather than LAD estimates are used to obtain MAD.

5.7f. Another type of test is the Wald test, which is based directly on the vector of estimates $\hat{\boldsymbol{\beta}}_2 = (\hat{\beta}_{q+1}, \ldots, \hat{\beta}_p)$ from the full model. If the hypothesis is true, then $\hat{\boldsymbol{\beta}}_2$ should be near $\mathbf{0}$. The Wald test statistic is $W = \boldsymbol{\beta}_2' A\hat{\boldsymbol{\beta}}_2/(p-q)\hat{\tau}^2$, where $A = X_2'X_2 - X_2'X_1(X_1'X_1)^{-1}X_1'X_2$ and X_1 and X_2 are, respectively, columns 1 through $q + 1$ and columns $q + 2$ through $p + 1$ of the matrix X and $\hat{\tau}^2$ is an estimate of the parameter τ^2 in Note 5.7d. The hypothesis is rejected if W is too large. See Schrader and Hettmansperger (1980, Section 3.3). The matrix A is chosen so that $\tau^2 A^{-1}$ is approximately equal to the variance-covariance matrix of $\hat{\boldsymbol{\beta}}_2$.

5.8a. Let us show that $\text{STR} = S_1 + 3\hat{\sigma}S_2$. By definition, $\text{STR} = \Sigma\rho(\hat{e}_i)$ and $S_1 + 3\hat{\sigma}S_2 = \Sigma f_i^2 + 3\hat{\sigma}\Sigma g_i = \Sigma(f_i^2 + 3\hat{\sigma}g_i)$, where $f_i = \max(-1.5\hat{\sigma}, \min(\hat{e}_i, 1.5\hat{\sigma}))$ and $g_i = \max(0, |\hat{e}_i| - 1.5\hat{\sigma})$. It suffices to show $\rho(\hat{e}_i) = f_i^2 + 3\hat{\sigma}g_i$. If $|\hat{e}_i| \le 1.5\hat{\sigma}$, then $\rho(\hat{e}_i) = \hat{e}_i^2$ and $f_i^2 + 3\hat{\sigma}g_i = \hat{e}_i^2 + 0 = \hat{e}_i^2$. If $|\hat{e}_i| > 1.5\hat{\sigma}$, then $\rho(\hat{e}_i) = 3\hat{\sigma}|\hat{e}_i| - 2.25\hat{\sigma}^2$ and $f_i^2 + 3\hat{\sigma}g_i = (1.5\hat{\sigma})^2 + 3\hat{\sigma}(|\hat{e}_i| - 1.5\hat{\sigma}) = 3\hat{\sigma}|\hat{e}_i| - 2.25\hat{\sigma}^2$.

5.8b. The numbers w_i are called weights. After the iterative procedure converges, the residuals $\hat{e}_i = y_i - (\hat{\beta}_0 + \hat{\beta}_1 x_{i1} + \cdots + \hat{\beta}_p x_{ip})$ satisfy the equations $\Sigma x_{ij} w_i \hat{e}_i = 0$, where $w_i = 2$ if $|\hat{e}_i| \le 1.5\hat{\sigma}$ and $w_i = 3\hat{\sigma}/|\hat{e}_i|$ if

$|\hat{e}_i| > 1.5\hat{\sigma}$. So the large residuals for which $|\hat{e}_i| > 1.5\hat{\sigma}$ are downweighted by weights $w_i < 2$. Thus the influence of outliers is reduced.

Additional Reading. To read more about M-regression, see the books by Huber (1981, Chapter 7), Hampel, Ronchetti, Rousseeuw, and Stahel (1986, Chapters 6 and 7), and Staudte and Sheather (1990, Chapter 7). The last book is more introductory than the other two.

REFERENCES

Becker, R. A., J. M. Chambers, and A. R. Wilks (1988). *The New S Language*. Wadsworth, Pacific Grove, CA.

Devore, J., and R. Peck (1986). *Statistics: The Exploration and Analysis of Data*. West, St. Paul, MN.

Hampel, F. R., E. M. Ronchetti, P. J. Rousseeuw, and W. A. Stahel (1986). *Robust Statistics: The Approach Based on Influence Functions*. Wiley, New York.

Holland, P. W., and R. E. Welsch (1977). Robust regression using iteratively reweighted least-squares. *Communications in Statistics A*, vol. 6, pp. 813–888.

Huber, P. (1964). Robust estimation of a location parameter. *Annals of Mathematical Statistics*, vol. 35, pp. 73–101.

Huber, P. (1981). *Robust Statistics*. Wiley, New York.

Marazzi, A. (1987). Solving bounded influence regression problems with ROBSYS. In: Y. Dodge (ed.), *Statistical Data Analysis Based on the L_1-Norm and Related Methods*. North-Holland, New York.

Rocke, D. M., and D. F. Shanno (1986). The scale problem in robust regression M-estimates. *Journal of Statistical Computing and Simulation*, vol. 24, pp. 47–69.

Rousseeuw, P. J., and A. J. Leroy (1987). *Robust Regression and Outlier Detection*. Wiley, New York.

SAS Institute Inc. (1990). *SAS/STAT User's Guide*, version 6, 4th ed., vol. 2. SAS Institute Inc., Cary, NC.

Schrader, R. M., and T. P. Hettmansperger (1980). Robust analysis of variance based on a likelihood criterion. *Biometrika*, vol. 67, pp. 93–101.

Shanno, D. F., and D. M. Rocke (1986). Numerical methods for robust regression: linear models. *SIAM Journal of Scientific and Statistical Computing*, vol. 7, pp. 86–97.

Staudte, R. G., and S. J. Sheather (1990). *Robust Estimation and Testing*. Wiley, New York.

CHAPTER 6

Nonparametric Regression

Ectasy is not reached by just anyone who starts to dance.
Dancing results from the soul's inner state; the inner
state of the soul does not result from dancing.

SOHRAVARDI: *Persian Philosopher (1155–1191)*

6.1 INTRODUCTION

A "parametric" statistical procedure is one whose justification depends on the assumption that the random errors in the data have a particular type of distribution. In particular, least-squares regression procedures are parametric in so far as they are optimal under the assumption that the errors have a normal distribution. A "robust" procedure, such as M-estimation in Chapter 5, is intended to perform reasonably well if the errors have a distribution that is not necessarily normal but "close" to normal. A "nonparametric" procedure is intended to perform reasonably well for almost any possible distribution of the errors. Many nonparametric procedures, including those described in this chapter, are based on the idea of using the ranks of numbers instead of the numbers themselves.

Perhaps the earliest occurrence of a statistical analysis based on ranks was one by Galton in 1876 to compare the heights of two types of plants. The development of nonparametric rank methods for regression data began in the 1960s and 1970s.

6.2 AN EXAMPLE OF SIMPLE REGRESSION

Physical measurements were taken on applicants to a police department in a city in the United States. Part of the data are shown in Table 6.1, which lists

Table 6.1 Forearm Length Data

Identification Number of the Applicant	Height (Y)	Forearm Length (X)
1	165.8	28.1
2	169.8	29.1
3	170.7	29.5
4	170.9	28.2
5	157.5	27.3
6	165.9	29.0
7	158.7	27.8
8	166.0	26.9
9	158.7	27.1
10	161.5	27.8
11	167.3	27.3
12	167.4	30.1
13	159.2	27.3
14	170.0	30.9
15	166.3	28.8
16	169.0	28.8
17	156.2	25.6
18	159.6	25.4
19	155.0	26.6
20	161.1	26.6
21	170.3	29.3
22	167.8	28.6
23	163.1	26.9
24	165.8	26.3
25	175.4	30.1
26	159.8	27.1
27	166.0	28.1
28	161.2	29.2
29	160.4	27.8
30	164.3	27.8
31	165.5	28.6
32	167.2	27.1
33	167.2	29.7

Source: Gunst and Mason (1980, p. 367).

the heights (in centimeters) and forearm lengths (also in centimeters) of 33 black female applicants.

First we make a plot consisting of 33 points, one point for each applicant, the coordinates of the point being x = length of forearm and y = height. This plot is shown in Figure 6.1. The points slope upward in a roughly linear manner, and so it is reasonable to try the simple linear regression model $Y = \alpha + \beta X + e$.

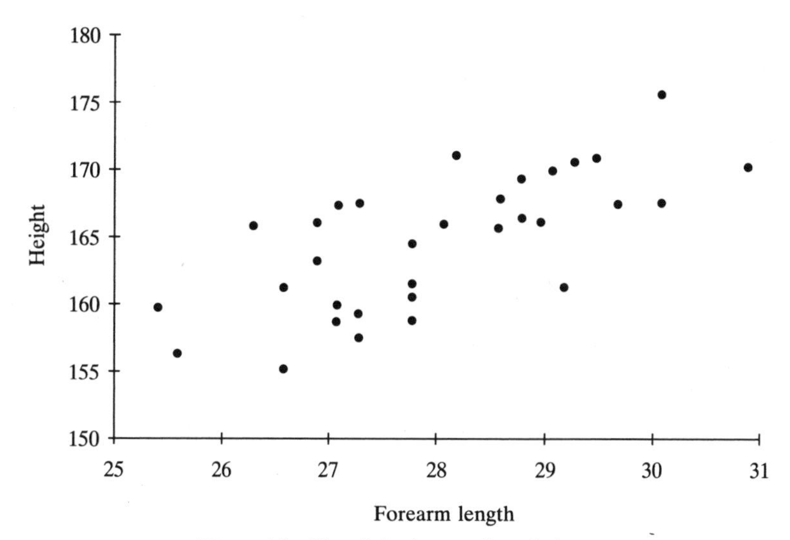

Figure 6.1 Plot of the forearm length data.

6.3 ESTIMATING THE REGRESSION LINE

An intuitively appealing approach to calculating the slope of a line that "fits" the data points is to calculate the slopes of all pairs of data points and then calculate some kind of average or median of these slopes. The slope of the line joining data points (x_i, y_i) and (x_j, y_j) is $b_{ij} = (y_i - y_j)/(x_i - x_j)$. (We ignore the pairs for which $x_i = x_j$ since then the pairwise slope is undefined.) The usual formula for the least-squares estimate of β is formula (3.2), but it can also be expressed as a weighted average of the pairwise slopes b_{ij}. Specifically, $\hat{\beta}_{\text{LS}} = \Sigma w_{ij} b_{ij}$, where $w_{ij} = (x_i - x_j)^2/\Sigma(x_i - x_j)^2$. (The two summations Σ in the preceding sentence are taken over the $n(n-1)/2$ pairs of integers i and j with $1 \leq i < j \leq n$.) Another reasonable estimate is the median of the pairwise slopes. The estimate presented in this section is a weighted median of the pairwise slopes.

Weighted Medians. Recall that the (unweighted) *median* of a list of numbers is obtained by putting the numbers in increasing order and selecting the number in the middle of the ordered list. For example, the median of the five numbers 203, 235, 47, 219, and 156 is obtained by putting them in the order 47, 156, 203, 219, 235, and selecting the middle number 203.

The weighted median of a list of numbers x_i with weights w_i is obtained as follows. First put the numbers x_i in increasing order. By changing the indices, we can arrange so that $x_1 \leq x_2 \leq \cdots \leq x_n$. The weights w_i should be nonnegative and should add to 1. Find the index k such that

$$\begin{aligned} w_1 + w_2 + \cdots + w_{k-1} &< 0.5 \\ w_1 + w_2 + \cdots + w_{k-1} + w_k &> 0.5 \end{aligned} \tag{6.1}$$

Then x_k is the *weighted median*. (Sometimes it happens that there is an index k such that $w_1 + w_2 + \cdots + w_{k-1} = 0.5$. Then $(x_{k-1} + x_k)/2$ is the weighted median.) As an example, suppose the numbers 203, 235, 47, 219, and 156 are given the weights 0.1, 0.1, 0.4, 0.1, and 0.3 respectively. In increasing order the numbers are 47, 156, 203, 219, 235. The weighted median is $x_2 = 156$ because $w_1 = 0.4 < 0.5$ and $w_1 + w_2 = 0.4 + 0.3 = 0.7 > 0.5$.

When all the weights are equal, that is, $w_i = 1/n$, then the weighted median is simply the ordinary median.

The slope β^* in (4.3), which is the slope of the line passing through the point (x_0, y_0) that minimizes the sum of absolute deviations from the data points, can be described as the weighted median of the slopes $b_i = (y_i - y_0)/(x_i - x_0)$ of the lines between the data points (x_i, y_i) and the given point (x_0, y_0), with each weight proportional to the x-distance $|x_i - x_0|$ between the two points.

Nonparametric Estimates of β and α. Let $\hat{\beta}$ be the weighted median of the pairwise slopes $b_{ij} = (y_i - y_j)/(x_i - x_j)$ with each weight proportional to the x-distance between the pair of points, that is, $w_{ij} = |x_i - x_j|/\Sigma|x_i - x_j|$. Let $\hat{\alpha}$ be the ordinary median of the differences $y_i - \hat{\beta}x_i$. This estimate of α makes sense if we note that in the model $y_i = \alpha + \beta x_i + e_i$, the differences $y_i - \beta x_i$ are centered around α.

The Forearm Length Data. For the 33 data points in Table 6.1 there are 528 (= 33(32)/2) distinct pairs of data points. For 18 of these pairs, the two data points have the same x-value, which means that the line between them does not have a well-defined slope. So we have 510 pairwise slopes. Each such slope $b_{ij} = (y_i - y_j)/(x_i - x_j)$ is assigned the weight $w_{ij} = |x_i - x_j|/802.2$, where 802.2 is obtained as the sum $\Sigma|x_i - x_j|$. For example, the slope between the data points for applicants 1 and 2 is $(165.8 - 169.8)/(28.1 - 29.1) = 4.000$ and its weight is $|28.1 - 29.1|/802.2 = 0.001247$.

We put the 510 slopes in increasing order and calculate the cumulative sums of the weights, as in Table 6.2. Now we find the first cumulative sum that exceeds 0.5. In the table we see that this is 0.504612. We estimate β by the corresponding slope, that is, $\hat{\beta} = 2.683$.

Next we calculate the 33 differences $y_i - \hat{\beta}x_i$. For example, the difference for applicant 1 is $165.8 - (2.683)(28.1) = 90.41$. We estimate α by the median of these differences, which is 89.71. The estimated regression line is $\hat{Y} = 89.71 + 2.683X$.

Table 6.2 Pairwise Slopes for the Forearm Length Data, in Increasing Order, and Their Weights

Slope b_{ij}	Weight w_{ij}	Cumulative Sum of Weights
−86.000	0.000125	0.000125
−48.500	0.000249	0.000374
−40.000	0.000249	0.000623
−36.500	0.000249	0.000873
−36.000	0.000374	0.001247
−31.000	0.000249	0.001496
⋮	⋮	⋮
2.604	0.006607	0.497632
2.667	0.001496	0.499127
2.667	0.000374	0.499501
2.683	0.005111	0.504612
2.684	0.002368	0.506981
2.684	0.004737	0.511718
⋮	⋮	⋮
37.500	0.000249	0.999252
39.000	0.000125	0.999377
43.000	0.000249	0.999626
49.000	0.000125	0.999751
51.000	0.000125	0.999875
91.000	0.000125	1.000000

A Description of $\hat{\beta}$ in Terms of Ranks. The estimates of α and β should be chosen so that the residuals $\hat{e}_i = y_i - (\hat{\alpha} + \hat{\beta}x_i)$ are "small". A sensible way to measure their smallness is by means of a weighted sum of the absolute values of the residuals, $\Sigma w_i|\hat{e}_i|$. The weights w_i should be nonnegative. In least-squares estimation, we choose $\hat{\alpha}$ and $\hat{\beta}$ to minimize the weighted sum with weights $w_i = |\hat{e}_i|$. In least-absolute-deviations estimation, we minimize the weighted sum with weights $w_i = 1$. An intermediate procedure, intermediate between weighting the residuals equally and weighting them according to their absolute values, would be to weight them according to the ranks of their absolute values, $w_i = \text{rank}(|\hat{e}_i|)$. (The ranking is done from smallest to largest, the smallest value of $|\hat{e}_i|$ being given the smallest rank, 1.) This would limit the influence of large residuals to a greater extent than least-squares estimation, since $\text{rank}(|\hat{e}_i|)$ can be no larger than n whereas $|\hat{e}_i|$ could be arbitrarily large, and to a lesser extent than least-absolute-deviations estimation. We will not use this procedure exactly but will use a similar procedure.

Rather than choose estimates that minimize $\Sigma\,\text{rank}(|\hat{e}_i|)|\hat{e}_i|$, we choose them to minimize

$$\Sigma\left[\text{rank}(\hat{e}_i) - \frac{n+1}{2}\right]\hat{e}_i \tag{6.2}$$

Both sums yield approximately the same estimates, at least when the population of errors has a symmetric distribution and when the sample size n is large. Sum (6.2) is expected to give better results when the distribution of the population of errors is not symmetric.

Below we will see that the nonparametric estimate $\hat{\beta}$, presented above as a weighted median, can also be characterized as the value of b that minimizes (6.2), where $\hat{e}_i = y_i - (a + bx_i)$. But first we verify the close relationship between the two sums.

Relationship Between the Rank-Weighted Sum of Absolute Residuals and Sum (6.2). If the distribution of the population of errors is approximately symmetric, then we can expect $\text{rank}(|\hat{e}_i|)$ to be approximately equal to $2|\text{rank}(\hat{e}_i) - \frac{1}{2}(n+1)|$. This is because $\frac{1}{2}(n+1)$ is the rank of the median residual, which can be expected to be near 0, and because the absolute values of the negative residuals should be approximately uniformly interspersed with the positive residuals. For example, suppose the residuals are $-23, -18, -11, 2, 16, 19, 29$. The values of $\text{rank}(|\hat{e}_i|)$ are, respectively, $6, 4, 2, 1, 3, 5, 7$. The values of $2|\text{rank}(\hat{e}_i) - \frac{1}{2}(n+1)|$ are, respectively, $6, 4, 2, 0, 2, 4, 6$. Note that dividing all the weights w_i by the same constant 2 will not affect the minimization of the weighted sum $\Sigma w_i|\hat{e}_i|$. So using the weights $w_i = \text{rank}(|\hat{e}_i|)$ is approximately equivalent to using the weights $w_i = |\text{rank}(\hat{e}_i) - \frac{1}{2}(n+1)|$.

If the median of the residuals is 0, then the negative residuals have rank less than $\frac{1}{2}(n+1)$ and the positive residuals have rank greater than $\frac{1}{2}(n+1)$. This implies that $\text{rank}(\hat{e}_i) - \frac{1}{2}(n+1)$ has the same sign as $\hat{e}_i$, so $|\text{rank}(\hat{e}_i) - \frac{1}{2}(n+1)|\,|\hat{e}_i| = [\text{rank}(\hat{e}_i) - \frac{1}{2}(n+1)]\hat{e}_i$. For example, suppose the residuals are $-23, -18, -11, 0, 16, 19, 29$. The values of $\text{rank}(\hat{e}_i) - \frac{1}{2}(n+1)$ for these residuals are $-3, -2, -1, 0, 1, 2, 3$. Note that $|-3|\,|-23| = (-3)(-23), \ldots, |0|\,|0| = (0)(0), \ldots, |3|\,|29| = (3)(29)$. Therefore, if the distribution of errors is approximately symmetric, in which case the median of the residuals can be expected to be near 0, then the weighted sum $\Sigma w_i|\hat{e}_i|$ with weights $w_i = |\text{rank}(\hat{e}_i) - \frac{1}{2}(n+1)|$ should be approximately equal to sum (6.2).

Equivalence of the Two Descriptions of $\hat{\beta}$. Let a and b denote candidates for $\hat{\alpha}$ and $\hat{\beta}$, and put $\hat{e}_i = y_i - (a + bx_i)$ in (6.2). First note that the value of (6.2) is unaffected by the value of a, and hence $\Sigma[\text{rank}(y_i - a - bx_i)) - \frac{1}{2}(n+1)](y_i - a - bx_i) = \Sigma[\text{rank}(y_i - bx_i) - \frac{1}{2}(n+1)](y_i - bx_i)$. This is because shifting all the residuals by the same amount a does not change their ranks, so that $\text{rank}(y_i - a - bx_i) = \text{rank}(y_i - bx_i)$, and because the sum of the n ranks must be $n(n+1)/2$, so that $\Sigma[\text{rank}(y_i - bx_i) - \frac{1}{2}(n+1)]a = 0$. Therefore the nonparametric estimate $\hat{\beta}$ is the value of b

that minimizes the sum

$$\sum \left[\text{rank}(y_i - bx_i) - \frac{n+1}{2} \right] (y_i - bx_i) \tag{6.3}$$

In particular, this shows that minimization of (6.2) can only determine an estimate of β and not of α. The estimate of α is calculated afterward by a different procedure.

Regard (6.3) as a function of b. We want to know why this function is minimized by choosing b to be the weighted median of the pairwise slopes b_{ij} with weights proportional to $|x_i - x_j|$.

For the forearm length data, the function is

$$\begin{aligned} &[\text{rank}(165.8 - 28.1b) - 17](165.8 - 28.1b) \\ &\quad + [\text{rank}(169.8 - 29.1b) - 17](169.8 - 29.1b) \\ &\quad + \cdots + [\text{rank}(167.2 - 29.7b) - 17](167.2 - 29.7b) \end{aligned} \tag{6.4}$$

The pairwise slopes b_{ij} and their weights are shown in Table 6.2. The graph of function (6.4) is shown in Figure 6.2. It consists of a series of line segments.

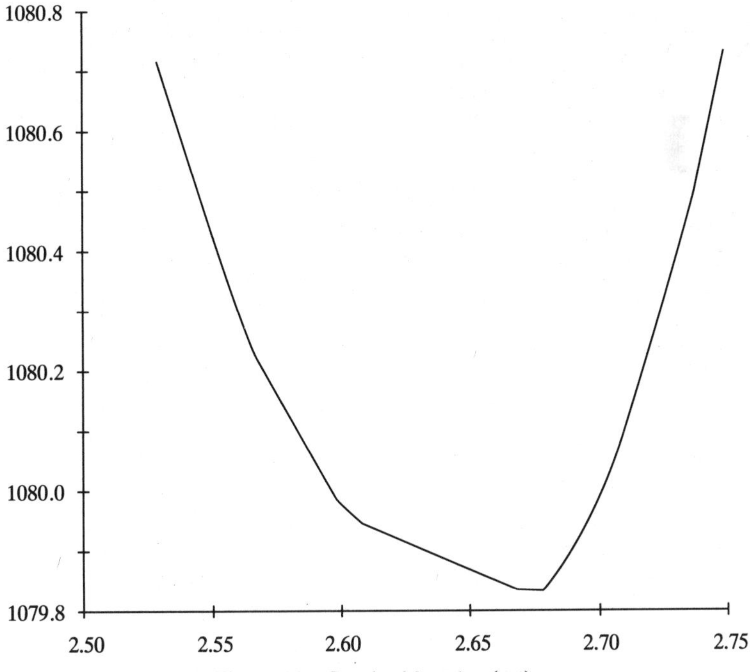

Figure 6.2 Graph of function (6.4).

More generally, the graph of function (6.3) consists of a series of line segments joined together at the points $b = b_{ij}$. In discussing this graph it is important to note that we are dealing with two different sorts of slopes: (1) the slope of function (6.3), which will tell us where the function has its minimum, and (2) the pairwise slopes b_{ij}, which are the points b at which slope (1) changes value.

The slope of (6.3) is the coefficient of b, which is $-\Sigma[r_i - \frac{1}{2}(n + 1)]x_i$, where $r_i = \text{rank}(y_i - bx_i)$. If b is close to b_{ij}, then $y_i - bx_i$ and $y_j - bx_j$ are close to each other (because $b_{ij} = (y_i - y_j)/(x_i - x_j)$) and hence their ranks differ only by 1. As b increases from slightly less than b_{ij} to slightly greater than b_{ij}, their ranks are interchanged, and hence r_i increases by 1 and r_j decreases by 1 (or vice versa, depending on whether $x_i < x_j$ or $x_i > x_j$). The resulting change in the slope of (6.3) is $|x_i - x_j|$.

As b varies from $-\infty$ to $+\infty$, the slope of (6.3) goes from $-\frac{1}{2}T$ to $\frac{1}{2}T$, where $T = \Sigma|x_i - x_j|$, increasing by $|x_i - x_j|$ at each point b_{ij}. At some point the slope must change from a negative value to a positive value, that is, the function stops decreasing and starts increasing. This of course is the minimizing value of b.

Put the pairwise slopes in increasing order and, for each b_{ij}, let T_{ij} be the cumulative sum of all $|x_f - x_g|$ for which $b_{fg} < b_{ij}$. The slope of (6.3) for b slightly less than b_{ij} is $-\frac{1}{2}T + T_{ij}$ and the slope for b slightly greater than b_{ij} is $-\frac{1}{2}T + T_{ij} + |x_i - x_j|$. The minimizing value of b is b_{km}, where $-\frac{1}{2}T + T_{km} < 0$ and $-\frac{1}{2}T + T_{km} + |x_k - x_m| > 0$. Dividing these conditions by T, we see that they are equivalent to (6.1), and hence b_{km} is the weighted median of the pairwise slopes b_{ij} with weights $w_{ij} = |x_i - x_j|/T$.

For the forearm length data, the slope of (6.4) in the interval from $b_{2,16} = 2.667$ to $b_{17,33} = 2.683$ is $-\frac{1}{2}T + T_{17,33} = -401.1 + 400.7 = -0.4 < 0$ ($401.1 = 802.2/2$, 802.2 is the sum of all the $|x_i - x_j|$, and 400.7 is the sum of the $|x_i - x_j|$ for which $b_{ij} < 2.683$). And the slope in the interval from $b_{17,33} = 2.683$ to $b_{4,24} = 2.684$ is $-\frac{1}{2}T + T_{17,33} + |x_{17} - x_{33}| = -0.4 + 4.1 = 3.7 > 0$. So the function stops decreasing and starts increasing at $\hat{\beta} = 2.683$.

6.4 TESTING $\beta = 0$

To test $\beta = 0$, we can use the test statistic

$$|t| = \frac{|U|}{\text{SD}(U)} \tag{6.5}$$

where

$$U = \sum \left[\text{rank}(y_i) - \frac{n+1}{2} \right] x_i$$

and

$$\text{SD}(U) = \sqrt{\frac{n(n+1)}{12} \sum (x_i - \bar{x})^2}$$

The approximate p-value of the test is calculated to be Prob$[|Z| \geq |t|]$, where Z is a random variable having a standard normal distribution.

Justification. The test statistics used in least-squares and least-absolute-deviations regression have the form $|t| = |\hat{\beta}|/\text{est.SD}(\hat{\beta})$ and are based on the fact that if $\beta = 0$, then $\hat{\beta}$ is likely to be near 0. This fact is true for any reasonable estimate of β, including the nonparametric estimate described in this chapter, but for the nonparametric $\hat{\beta}$ it is difficult to obtain a good estimate of SD($\hat{\beta}$). Test statistic (6.5) is based on the fact that if $\beta = 0$, then U is likely to be near 0. Therefore, if $|t|$ is large, that is, if U is far from 0 relative to the magnitude of SD(U), then we conclude that $\beta \neq 0$.

The reason we expect U to be near 0 if $\beta = 0$ is that then its expectation is 0. When $\beta = 0$, then $y_i = \alpha + e_i$ and the observations $y_1, y_2, \ldots, y_n$ can be regarded as having been independently selected from the same population. This implies that the expectation of rank(y_i) is the same for all i. Since the ranks must add to $n(n+1)/2$, the expectation of each rank(y_i) must be $\frac{1}{2}(n+1)$. Hence $[\text{rank}(y_i) - \frac{1}{2}(n+1)]$ has expectation 0 for all i, and so U has expectation 0.

Another reason for expecting U to be near 0 when $\beta = 0$ is the following. If $\beta = 0$, then the nonparametric estimate $\hat{\beta}$ should be near 0, which means that the minimum of function (6.3) should be near 0. Near the minimum of the function its slope is near 0. Therefore, if $\beta = 0$, the slope of (6.3) should be near 0 at $b = 0$. Note that $-U$ is the slope of function (6.3) in the interval containing $b = 0$.

The p-Value. When the sample size n is large, it is known that $t = U/\text{SD}(U)$ has approximately a standard normal distribution if the null hypothesis $\beta = 0$ is true. So we have more confidence in the accuracy of the p-value for larger sample sizes. To be on the conservative side, one might calculate the p-value using the t distribution with $n - 2$ degrees of freedom instead of the standard normal distribution. This would be conservative in that, when the null hypothesis is true, that is, when the y-values have no relationship with the x-values, we would be less likely to falsely conclude that there is a significant relationship.

Similarity to the Least-Squares Test. We can write (6.5) in a form that is similar to the least-squares test statistic. Recall that the least-squares test uses $t_{\text{LS}} = \hat{\beta}_{\text{LS}}/\text{est.SD}(\hat{\beta}_{\text{LS}})$, where $\hat{\beta}_{\text{LS}} = \Sigma(x_i - \bar{x})(y_i - \bar{y})/\Sigma(x_i - \bar{x})^2$ and $\text{est.SD}(\hat{\beta}_{\text{LS}}) = \hat{\sigma}/\sqrt{\Sigma(x_i - \bar{x})^2}$. Altogether,

$$t_{\text{LS}} = \frac{\Sigma(x_i - \bar{x})(y_i - \bar{y})}{\hat{\sigma}\sqrt{\Sigma(x_i - \bar{x})^2}}$$

The nonparametric test uses $t_{\text{NP}} = U/\text{SD}(U)$. Let r_i denote rank(y_i). The average rank is $\bar{r} = \frac{1}{2}(n + 1)$, and so $U = \Sigma(r_i - \bar{r})x_i$. Using the fact that $\Sigma(r_i - \bar{r}) = 0$, we can write $U = \Sigma(x_i - \bar{x})(r_i - \bar{r})$, and hence

$$t_{\text{NP}} = \frac{\Sigma(x_i - \bar{x})(r_i - \bar{r})}{\sqrt{\dfrac{n(n+1)}{12}}\sqrt{\Sigma(x_i - \bar{x})^2}}$$

This is like t_{LS} with r_i in place of y_i and $\sqrt{n(n+1)/12}$ in place of $\hat{\sigma}$. The similarity goes further because $\hat{\sigma}$ is an estimate of the SD of the y_i's and $\sqrt{n(n+1)/12}$ can be shown to be equal to the sample SD of the r_i's. (Note that this analogy between the least-squares and nonparametric rank-based procedures for testing β does not extend to estimation of β. In particular, we are not suggesting that $\Sigma(x_i - \bar{x})(r_i - \bar{r})/\Sigma(x_i - \bar{x})^2$ is a good estimate of β).

This gives another way to obtain t_{NP}, as $\sqrt{n - 1}$ times the correlation coefficient between the x_i's and the r_i's.

Ties. If ties occur among the observations y_i, it is not clear how to obtain their ranks, which are needed to calculate the test statistic $|t|$. We can use midranks. For example, in the forearm length data, applicants 1 and 24 have the same y-value 165.8. If their heights had been, say, 165.79 and 165.81, then their ranks would have been 16 and 17. But they both have the same height 165.8, so we assign both applicants the *midrank* 16.5.

The theory underlying test statistic (6.5) assumes that the population of errors has a continuous distribution, which guarantees that no ties occur. The theory does not strictly pertain to data with ties, but if there are only a few ties, the test should produce a p-value that is adequately accurate.

The Forearm Length Data. Let us apply a nonparametric test to the data in Table 6.1 to see whether forearm length has a significant relationship with height. We first find the ranks of the 33 heights. This can be done by putting the heights in increasing order. The shortest height is 155.0, so its rank is 1; the second shortest height is 156.2, so its rank is 2; and so on. Midranks are used for ties that occur for four pairs of applicants. An additional column for rank(y_i) could be appended to Table 6.1: $16.5, 28, \ldots, 22.5$. Then $U = (16.5 - 17)(28.1) + (28 - 17)(29.1) + \cdots + (22.5 - 17)(29.7) = 286.4$. Next calculate $n(n + 1)/12 = 93.5$, $\Sigma(x_i - \bar{x})^2 = 55.56$, and $SD(U) = \sqrt{(93.5)(55.56)} = 72.08$. The test statistic is $|t| = |286.4|/72.08 = 3.973$. Looking in the t table in the Appendix at the row for 31 (= 33 − 2) degrees of freedom, we see that the p-value is less than 0.001. This indicates that forearm length has a significant relationship with height.

6.5 AN EXAMPLE OF MULTIPLE REGRESSION

A regression model can be applied to the data in Table 6.3 to see how the amount of money that a state spends on education is related to other characteristics of the state. For each of the 50 states in the United States the table lists the money spent on public education relative to the population of

Table 6.3 Education Expenditure Data

State	Expenditure on Education (Y)	Personal Income (X_1)	Youth Percentage (X_2)	Urban Percentage (X_3)
ME	235	3944	32.5	50.8
NH	231	4578	32.3	56.4
VT	270	4011	32.8	32.2
MA	261	5233	30.5	84.6
RI	300	4780	30.3	87.1
CT	317	5889	30.7	77.4
NY	387	5663	30.1	85.6
NJ	285	5759	31.0	88.9
PA	300	4894	30.0	71.5
OH	221	5012	32.4	75.3
IN	264	4908	32.9	64.9
IL	308	5753	32.0	83.0
MI	379	5439	33.7	73.8
WI	342	4634	32.8	65.9
MN	378	4921	33.0	66.4
IA	232	4869	31.8	57.2
MO	231	4672	30.9	70.1
ND	246	4782	33.3	44.3
SD	230	4296	33.0	44.6
NB	268	4827	31.8	61.5
KS	337	5057	30.4	66.1
DE	344	5540	32.8	72.2

Table 6.3 ***(Continued)***

State	Expenditure on Education (Y)	Personal Income (X_1)	Youth Percentage (X_2)	Urban Percentage (X_3)
MD	330	5331	32.3	76.6
VA	261	4715	31.7	63.1
WV	214	3828	31.0	39.0
NC	245	4120	32.1	45.0
SC	233	3817	34.2	47.6
GA	250	4243	33.9	60.3
FL	243	4647	28.7	80.5
KY	216	3967	32.5	52.3
TN	212	3946	31.5	58.8
AL	208	3724	33.2	58.4
MS	215	3448	35.8	44.5
AR	221	3680	32.0	50.0
LA	244	3825	35.5	66.1
OK	234	4189	30.6	68.0
TX	269	4336	33.5	79.7
MT	302	4418	33.5	53.4
ID	268	4323	34.4	54.1
WY	323	4813	33.1	60.5
CO	304	5046	32.4	78.5
NM	317	3764	36.6	69.8
AZ	332	4504	34.0	79.6
UT	315	4005	37.8	80.4
NV	291	5560	33.0	80.9
WA	312	4989	31.3	72.6
OR	316	4697	30.5	67.1
CA	332	5438	30.7	90.9
AK	311	5309	33.3	83.1
HI	546	5613	38.6	48.8

Source: Chatterjee and Price (1977, p. 99).

the state (in dollars per resident) in the year 1975, personal income (in dollars per resident) in 1973, the percentage of residents under the age of 18 in 1974, and the percentage of residents living in urban areas in 1970. The linear regression model is $y = \beta_0 + \beta_1 X_1 + \beta_2 X_2 + \beta_3 X_3 + e$.

Looking at the three plots of Y versus the explanatory variables X_1, X_2, and X_3, we see no clear departure form the model's assumptions that the regression function is linear and the errors have a common variance.

6.6 ESTIMATING THE REGRESSION COEFFICIENTS

Earlier in this chapter, the nonparametric estimate of the slope of a regression line in simple regression is described in two ways: as a weighted median of pairwise slopes and as the minimizing value for the sum (6.3). The second description is easily generalized to multiple regression. The nonparametric estimates $\hat{\beta}_1, \ldots, \hat{\beta}_p$ are the values of $b_1, \ldots, b_p$ that minimize

$$\sum \left[\text{rank}\big(y_i - (b_1 x_{i1} + \cdots + b_p x_{ip}) \big) - \frac{n+1}{2} \right] \times \big(y_i - (b_1 x_{i1} + \cdots + b_p x_{ip}) \big) \tag{6.6}$$

Then the nonparametric estimate $\hat{\beta}_0$ is obtained as the median of the differences $y_i - (\hat{\beta}_1 x_{i1} + \cdots + \hat{\beta}_p x_{ip})$.

The justifications of the estimates $\hat{\beta}_1, \ldots, \hat{\beta}_p$ and $\hat{\beta}_0$ are the same as in Section 6.3. Note that (6.6) is the same as (6.2) with $\hat{e}_i = y_i - (b_1 x_{i1} + \cdots + b_p x_{ip})$. The argument preceding (6.2) shows that minimizing (6.6) is approximately equivalent to minimizing a certain weighted sum of the absolute residuals. The estimate $\hat{\beta}_0$ is based on the fact that the differences $y_i - (\beta_1 x_{i1} + \cdots + \beta_p x_{ip})$ are equal to $\beta_0 + e_i$ and hence are centered around β_0.

An Algorithm for Minimizing (6.6). It is convenient to give function (6.6) a name, say, g, and to express it in vector notation:

$$g(\boldsymbol{b}) = \sum \left[\text{rank}(y_i - \boldsymbol{b}'\boldsymbol{x}_i) - \frac{n+1}{2} \right] (y_i - \boldsymbol{b}'\boldsymbol{x}_i) \tag{6.7}$$

where $\boldsymbol{b} = (b_1, \ldots, b_p)$ and $\boldsymbol{x}_i = (x_{i1}, \ldots, x_{ip})$. We want to find a vector $\boldsymbol{b}$ that minimizes $g(\boldsymbol{b})$. This can be done by an iterative procedure which, starting with the vector of least-squares estimates, finds vectors that give smaller and smaller values of the function g.

Given the current vector $\boldsymbol{b}^0$, a better vector is obtained by forming $\boldsymbol{b}^* = \boldsymbol{b}^0 + t^*\boldsymbol{d}$, where t^* and $\boldsymbol{d}$ are calculated as follows. Let $z_i = y_i - (\boldsymbol{b}^0)'\boldsymbol{x}_i$, $u_i^0 = \text{rank}(z_i) - \frac{1}{2}(n+1)$, $\boldsymbol{u}^0 =$ the $n \times 1$ vector with entries u_i^0, and $\boldsymbol{X}_c =$ the $n \times p$ matrix with entries $x_{ij} - \bar{x}_j$. Then $\boldsymbol{d} = (\boldsymbol{X}_c'\boldsymbol{X}_c)^{-1}\boldsymbol{X}_c'\boldsymbol{u}^0$. Let $w_i = \boldsymbol{d}'\boldsymbol{x}_i$ and let t^* be the weighted median of the ratios $(z_i - z_j)/(w_i - w_j)$ with weights $|w_i - w_j| / \Sigma |w_i - w_j|$. This completes one iteration.

Now $\boldsymbol{b}^*$ becomes the current vector and the iteration is repeated.

Justification of the Algorithm. Given the current vector $\boldsymbol{b}^0$, an improved vector $\boldsymbol{b}^*$ with $g(\boldsymbol{b}^*) < g(\boldsymbol{b}^0)$ is found in two steps. First we find a direction in which the value of g decreases; that is, we find a vector $\boldsymbol{d}$ such that $g(\boldsymbol{b}^0 + t\boldsymbol{d})$ decreases as t increases from 0. Then we find the value of t, say, t^*, that minimizes $g(\boldsymbol{b}^0 + t\boldsymbol{d})$. The vector $\boldsymbol{b}^* = \boldsymbol{b}^0 + t^*\boldsymbol{d}$ has a smaller value of g than $\boldsymbol{b}^0$ does.

First we want to find a vector $\boldsymbol{d}$ such that $g(\boldsymbol{b}^0 + t\boldsymbol{d})$ decreases as t increases from 0. Such a decrease occurs if and only if the derivative of $g(\boldsymbol{b}^0 + t\boldsymbol{d})$ at $t = 0$ is negative. This derivatives is $(\nabla g(\boldsymbol{b}^0))'\boldsymbol{d}$, where $\nabla g(\boldsymbol{b})$ denotes the vector of partial derivatives of $g(\boldsymbol{b})$ with respect to b_j for $j = 1, \ldots, p$. (For the moment, let us ignore the fact that these partial derivatives are not well-defined at some points $\boldsymbol{b}$.) If we are lucky enough that the current vector of estimates $\boldsymbol{b}^0$ satisfies $\nabla g(\boldsymbol{b}^0) = \boldsymbol{0}$, that is, if all the partial derivatives of $g(\boldsymbol{b})$ are 0 at $\boldsymbol{b} = \boldsymbol{b}^0$, then this implies that the minimum of $g(\boldsymbol{b})$ occurs at $\boldsymbol{b} = \boldsymbol{b}^0$, and so the algorithm has found the vector of nonparametric estimates. Otherwise, we can make the derivative of $g(\boldsymbol{b}^0 + t\boldsymbol{d})$ negative at $t = 0$ by choosing the direction vector to be $\boldsymbol{d} = -\nabla g(\boldsymbol{b}^0)$. But for reasons based on the variances and covariances of the regression estimates, the choice $\boldsymbol{d} = -(\boldsymbol{X}_c'\boldsymbol{X}_c)^{-1}\nabla g(\boldsymbol{b}^0)$ has been suggested.

The partial derivative of $g(\boldsymbol{b})$ with respect to b_j is $-\Sigma[\text{rank}(y_i - \boldsymbol{b}'\boldsymbol{x}_i) - \frac{1}{2}(n + 1)]x_{ij}$. Using the fact that $\Sigma[\text{rank}(y_i - \boldsymbol{b}'\boldsymbol{x}_i) - \frac{1}{2}(n + 1)] = 0$, the partial derivative can also be written as $-\Sigma[\text{rank}(y_i - \boldsymbol{b}'\boldsymbol{x}_i) - \frac{1}{2}(n + 1)](x_{ij} - \bar{x}_j)$. Hence $\nabla g(\boldsymbol{b}^0) = -\boldsymbol{X}_c'\boldsymbol{u}^0$, and so the suggested direction vector can be expressed as $\boldsymbol{d} = (\boldsymbol{X}_c'\boldsymbol{X}_c)^{-1}\boldsymbol{X}_c'\boldsymbol{u}^0$.

For a given direction vector $\boldsymbol{d}$, we search along this direction for the value $t = t^*$ that minimizes $g(\boldsymbol{b}^0 + t\boldsymbol{d})$. From (6.7) we have $g(\boldsymbol{b}^0 + t\boldsymbol{d}) = \Sigma[\text{rank}(f_i) - \frac{1}{2}(n + 1)]f_i$, where $f_i = y_i - (\boldsymbol{b}^0 + t\boldsymbol{d})'\boldsymbol{x}_i$. Note that $f_i = z_i - tw_i$. Therefore finding the value of t that minimizes $g(\boldsymbol{b}^0 + t\boldsymbol{d})$ is the same as finding the value of t that minimizes (6.3) with b replaced by t, y_i by z_i, and x_i by w_i. In other words, t^* is the nonparametric estimate of the slope of the regression line for the model $Z = \alpha + \beta W + e$. As seen in Section 6.3, t^* is the weighted median of the ratios $(z_i - z_j)/(w_i - w_j)$ with weights proportional to $|w_i - w_j|$.

When the Derivatives Are Not Well-Defined. When the partial derivative of $g(\boldsymbol{b})$ with respect to b_j is well-defined, it is equal to $-\Sigma[\text{rank}(y_i - \boldsymbol{b}'\boldsymbol{x}_i) - \frac{1}{2}(n + 1)]x_{ij}$. The derivative is not well-defined at those points $\boldsymbol{b}$ at which some of the residuals $y_i - \boldsymbol{b}'\boldsymbol{x}_i$ are equal because then their ranks are not well-defined. At such points we can use midranks to calculate an approximate derivative. (See the subsection on ties in Section 6.4.)

When $p = 1$, the points of nondifferentiability are the points $b_{ij} = (y_i - y_j)/(x_i - x_j)$. Note that at $b = b_{ij}$ the residuals $y_i - bx_i$ and $y_j - bx_j$ are equal. Recall that the graph of function (6.3) is a series of line segments. The points b_{ij} are the "corners" where the line segments meet.

The Education Expenditure Data. Let us use the algorithm described above to obtain nonparametric estimates of the regression coefficients for the education expenditure data. The procedure to estimate β_1, β_2, and β_3 is iterative, starting with the vector of least-squares estimates $\boldsymbol{b}^0 = (0.07239, 15.52, -0.04269)$.

To improve $\boldsymbol{b}^0$ we first find a good direction vector. Calculate the 50 differences $y_i - (\boldsymbol{b}^0)'\boldsymbol{x}_i$. For example, the difference for Maine is $235 - [(0.07239)(3944) + (15.52)(32.5) + (-0.04269)(50.8)] = -552.7$. Ranking the 50 differences, we obtain the entries $u_i^0 = \text{rank}(y_i - (\boldsymbol{b}^0)'\boldsymbol{x}_i) - 25.5$ of the vector $\boldsymbol{u}^0$, which is used to calculate a good direction vector $\boldsymbol{d} = (\boldsymbol{X}_c'\boldsymbol{X}_c)^{-1}\boldsymbol{X}_c'\boldsymbol{u}^0$. The 50×3 matrix $\boldsymbol{X}_c$ is obtained from the last three columns of Table 6.3 by subtracting $\bar{x}_1 = 4675$, $\bar{x}_2 = 32.57$, and $\bar{x}_3 = 65.78$, respectively, from the three columns. We obtain $\boldsymbol{d} = (-0.002505, -0.5157, 0.08182)$.

Next we minimize $g(\boldsymbol{b}^0 + t\boldsymbol{d})$. For this we need the quantities $z_i = y_i - (\boldsymbol{b}^0)'\boldsymbol{x}_i$ and $w_i = \boldsymbol{d}'\boldsymbol{x}_i$. The z_i have already been calculated in the process of obtaining $\boldsymbol{d}$. We now calculate $w_1 = \boldsymbol{d}'\boldsymbol{x}_1 = (-0.002505)(3944) + (-0.5157)(32.5) + (0.08182)(50.8) = -22.48$, and so on. The minimizing value of t is the weighted median of the ratios $(z_i - z_j)/(w_i - w_j)$ with weights $|w_i - w_j|/\Sigma|w_i - w_j|$. The weighted median can be found by constructing a table similar to Table 6.2. Put the ratios in increasing order and list them in the first column of the table. List the corresponding weights in the second column and calculate the cumulative sum of the weights in the third column. Find the cumulative sum that first exceeds 0.5. The corresponding ratio is $t^* = 5.346$. Therefore the improved vector of estimates is $\boldsymbol{b}^* = \boldsymbol{b}^0 + t^*\boldsymbol{d} = (0.07239 + (5.346)(-0.002505), 15.52 + (5.346)(-0.5157), -0.04269 + (5.346)(0.08182)) = (0.05900, 12.76, 0.3947)$. This completes the first iteration.

Relabeling $\boldsymbol{b}^*$ as $\boldsymbol{b}^0$, we repeat the same procedure to further improve the vector of estimates. We calculate the differences $z_i = y_i - (\boldsymbol{b}^0)'\boldsymbol{x}_i$, rank them to obtain the entries of $\boldsymbol{u}^0$, and then calculate $\boldsymbol{d} = (\boldsymbol{X}_c'\boldsymbol{X}_c)^{-1}\boldsymbol{X}_c'\boldsymbol{u}^0 = (-0.00002842, -0.09464, -0.02485)$. Next we calculate $w_i = \boldsymbol{d}'\boldsymbol{x}_i$ and find the weighted median of the ratios $(z_i - z_j)/(w_i - w_j)$ with weights $|w_i - w_j|/\Sigma|w_i - w_j|$. The weighted median is $t^* = 2.533$ and the improved vector is $\boldsymbol{b}^0 + t^*\boldsymbol{d} = (0.05900 + (2.533)(-0.00002842), 12.76 + (2.533)(-0.09464), 0.3947 + (2.533)(-0.02485) = (0.05892, 12.52, 0.3317)$. This completes the second iteration. At the 15th iteration, the relative changes in all three estimates, b_1^0, b_2^0, and b_3^0, are less than 10^{-8}, so we can feel confident that, to at least four significant digits, these are the values of the nonparametric regression estimates: $\hat{\beta}_1 = 0.05868$, $\hat{\beta}_2 = 12.54$, and $\hat{\beta}_3 = 0.3372$.

We obtain the nonparametric estimate $\hat{\beta}_0$ as the median of the 50 differences $y_i - (0.05868x_{i1} + 12.54x_{i2} + 0.3372x_{i3})$, which is -422.9. So the estimated regression equation is $\hat{Y} = -422.9 + 0.05868X_1 + 12.54X_2 + 0.3372X_3$.

Estimation of β_0 When the Error Distribution Is Symmetric. We have estimated β_0 by the median of the differences $d_i = y_i - (\hat{\beta}_1 x_{i1} + \cdots + \hat{\beta}_p x_{ip})$. If the distribution of the population of errors can be assumed to be symmetric, then a better estimate is the median of the pairwise averages $(d_i + d_j)/2$, $1 \leq i \leq j \leq n$. For the education expenditure data, the two estimates of β_0 are almost the same, both being equal to -422.9 to four significant digits.

6.7 TESTING $\beta_{q+1} = \cdots = \beta_p = 0$

The nonparametric test of $\beta_{q+1} = \cdots = \beta_p = 0$ is analogous to the least-squares test. Recall that the least-squares test statistic is

$$F_{\text{LS}} = \frac{\text{SSR}_{\text{reduced}} - \text{SSR}_{\text{full}}}{(p-q)\hat{\sigma}^2_{\text{LS}}}$$

where SSR stands for the sum of squared residuals, $\text{SSR} = \Sigma \hat{e}_i^2$.

Description of the Test. Let us introduce the notation SRWR for the sum of rank-weighted residuals, $\text{SRWR} = \Sigma[\text{rank}(\hat{e}_i) - \frac{1}{2}(n+1)]\hat{e}_i$. The nonparametric test statistic is

$$F_{\text{NP}} = \frac{\text{SRWR}_{\text{reduced}} - \text{SRWR}_{\text{full}}}{(p-q)c\hat{\tau}} \tag{6.8}$$

where $c = (n+1)/\sqrt{48}$ and $\hat{\tau}$ is given by formula (6.9) below. The residuals in $\text{SRWR}_{\text{reduced}}$ and $\text{SRWR}_{\text{full}}$ are calculated by applying the nonparametric regression method to the reduced model $Y = \beta_0 + \beta_1 X_1 + \cdots + \beta_q X_q + e$ and the full model $Y = \beta_0 + \beta_1 X_1 + \cdots + \beta_p X_p + e$, respectively. Note that SRWR is the minimum value of (6.7); that is, $\text{SRWR} = g(\hat{\boldsymbol{\beta}})$.

To calculate $\hat{\tau}$, take the residuals from the full model and form the pairwise averages, $A_{ij} = (\hat{e}_i + \hat{e}_j)/2$ for $1 \leq i \leq j \leq n$. Put these $N = n(n+1)/2$ numbers in increasing order: $A_{(1)} \leq A_{(2)} \leq \cdots \leq A_{(N)}$. Let $a = n(n+1)/4$, $b = \sqrt{n(n+1)(2n+1)/24}$, $k_1 =$ the closest integer to $\frac{1}{2} + a - (1.645)b$, $k_2 =$ the closest integer to $\frac{1}{2} + a + (1.645)b$, and $f = \sqrt{n/[n-(p+1)]}$. Define

$$\hat{\tau} = f\frac{\sqrt{n}\left[A_{(k_2)} - A_{(k_1)}\right]}{2(1.645)} \tag{6.9}$$

Just as in the least-squares test, an approximate p-value of the nonparametric test is calculated to be $\text{Prob}[F \geq F_{\text{NP}}]$, where F denotes a random variable having an F distribution with $p - q$ and $n - p - 1$ degrees of freedom.

Justification of (6.8). The sum SRWR plays the same role in nonparametric estimation that the sum SSR plays in least-squares estimation. The sum SSR measures how well the least-squares estimated regression equation fits the data. Similarly, the sum SRWR measures how well the nonparametric estimated regression equation fits the data. If the null hypothesis is false, then the reduced model is not the true model, and so the regression equation estimated using the reduced model tends to fit the data poorly. Thus $\text{SRWR}_{\text{reduced}}$ tends to be larger when the null hypothesis is false than when the null hypothesis is true.

The nonparametric test statistic is formed from $\text{SRWR}_{\text{reduced}}$ by subtracting $\text{SRWR}_{\text{full}}$ and dividing by $(p - q)c\hat{\tau}$. This serves to standardize the test statistic so that F_{NP} tends to be near 1 when the null hypothesis is true and to be significantly larger than 1 when the null hypothesis is false.

The Parameter τ. The denominator of test statistic (6.8) is $((p - q)(n + 1) / \sqrt{48})\hat{\tau}$. The estimate $\hat{\tau}$ estimates a parameter τ which plays a role in nonparametric regression similar to the role of σ in least-squares regression.

The vector of least-squares estimates, $\hat{\boldsymbol{\beta}}_{\text{LS}}$, has its variance–covariance matrix equal to $\sigma^2(\boldsymbol{X}'\boldsymbol{X})^{-1}$, where $\boldsymbol{X}$ is the matrix of explanatory variables with a vector of 1's added as the first column. In nonparametric regression the primary part of the estimation procedure concerns $\beta_1, \ldots, \beta_p$; estimation of β_0 is a separate part. Let $\boldsymbol{\delta} = (\beta_1, \ldots, \beta_p)$. The variance–covariance matrix of $\hat{\boldsymbol{\delta}}_{\text{LS}}$ is $\sigma^2\boldsymbol{W}$, where $\boldsymbol{W}$ is obtained from $(\boldsymbol{X}'\boldsymbol{X})^{-1}$ by omitting the first row and first column.

The vector of nonparametric estimates, $\hat{\boldsymbol{\delta}}_{\text{NP}}$, has its variance–covariance matrix approximately equal to $\tau^2\boldsymbol{W}$. (The approximation is best for large sample sizes.) In particular, the standard deviation of $\hat{\beta}_{j,\text{NP}}$ is approximately equal to τ/σ times the standard deviation of $\hat{\beta}_{j,\text{LS}}$. Hence, whether the nonparametric estimates are more accurate than the least-squares estimates depends on whether $\tau < \sigma$.

Suppose the distribution of the population of errors can be assumed to be symmetric. Then a good choice for $\hat{\beta}_{0,\text{NP}}$ is the median of the pairwise averages $(d_i + d_j)/2$, where $d_i = y_i - (\hat{\beta}_1 x_{i1} + \cdots + \hat{\beta}_p x_{ip})$. In this case, the variance–covariance matrix of $\hat{\boldsymbol{\beta}}_{\text{NP}}$ is approximately $\tau^2(\boldsymbol{X}'\boldsymbol{X})^{-1}$.

Both σ and τ can be regarded as measures of variability in the distribution of errors. The parameter σ is the standard deviation. The parameter τ can be described in terms of the distribution of differences between pairs of errors: $\tau = 1/(\sqrt{12}\,\gamma)$, where γ is the probability density of the error differences at 0. If the distribution of errors is highly variable, then any two independently selected errors will tend to be far from one another, so their difference will tend to be far from 0, which implies that the probability density of error differences near 0 is low; that is, γ is small, and hence τ is large.

Roughly speaking, the more "spread out" the distribution of errors is, the larger both σ and τ are. But the exact ratio τ/σ depends on the shape of the distribution of the errors. If the errors have a normal distribution, then $\tau/\sigma = 1.023 > 1$, and so, at least for large sample sizes, the nonparametric regression estimates are slightly less accurate than the least-squares regression estimates. If the errors have a uniform distribution, then $\tau/\sigma = 1$. If the errors have a Laplace distribution, then $\tau/\sigma = 0.816 < 1$.

Justification of (6.9). The estimate of τ in formula (6.9) has been found to perform well when the errors have a symmetric distribution. Of course it would be preferable to have an estimate that also performed well for asymmetric error distributions, but the candidates that have been proposed so far are either unsatisfactory or not yet sufficiently investigated.

The residuals $\hat{e}_i$ behave roughly the same as the true errors e_i, especially for large sample sizes. It is convenient to present the following justification as if $\hat{e}_i = e_i$.

The argument is based on a comparison of two confidence intervals for the mean of the population of errors. Of course the regression model assumes that this mean is 0, and so we are not actually interested in the confidence intervals themselves but only in their lengths. To make it easier to think in terms of confidence intervals, pretend we do not know the mean μ of the population from which the e_i are drawn. We will calculate two 90%-confidence intervals for μ. The two intervals tend to have about the same lengths. Equating the two lengths yields formula (6.9).

In constructing the first confidence interval, we assume that the distribution of the population of errors is symmetric. Then, as indicated above in the subsection on the parameter τ (with $p = 0$ and $\beta_0 = \mu$), a good nonparametric estimate of μ is the median of the $n(n+1)/2$ pairwise averages $(e_i + e_j)/2$. Denote this estimate by $\hat{\mu}$. The mean of $\hat{\mu}$ is approximately μ and its standard deviation is approximately $\tau/\sqrt{n}$ (when $p = 0$, then X is a column of 1's and so $\tau^2(X'X)^{-1} = \tau^2/n$). For large sample sizes n, the distribution of $\hat{\mu}$ is approximately normal. Therefore, if $\hat{\tau}$ is an estimate of τ, an approximate 90%-confidence interval for μ is $\hat{\mu} \pm (1.645)\hat{\tau}/\sqrt{n}$.

Next, it can be shown that the interval from $A_{(k_1)}$ to $A_{(k_2)}$ is also a 90%-confidence interval for μ. The lengths of the two intervals, at least for large sample sizes n, are approximately the same, that is, $2(1.645)\hat{\tau}/\sqrt{n} \approx$

$A_{(k_2)} - A_{(k_1)}$. Choosing $\hat{\tau}$ to make this an equality yields (6.9) without f. The factor f is introduced to improve the accuracy of the p-value of the test when the sample size n is small. The factor has little effect for large sample sizes since then $f \approx 1$.

The Education Expenditure Data. From the data in Table 6.3 we have estimated a regression equation that expresses the education expenditure of a state as a function of three explanatory variables. Now let us test whether the only significant explanatory information contained in these three variables is actually contained in the single variable of personal income. In other words, let us test whether $\beta_2 = \beta_3 = 0$.

We want to calculate the test statistic F_{NP}. In Section 6.6 we estimated the regression coefficients for the full model. These are used to calculate the residuals $\hat{e}_i = y_i - (-422.9 + 0.05868x_{i1} + 12.54x_{i2} + 0.3372x_{i3})$. We then rank the residuals and calculate $\text{SRWR}_{\text{full}} = \Sigma[\text{rank}(\hat{e}_i) - 25.5]\hat{e}_i = 27120$. For the reduced model without X_2 and X_3, the estimated regression coefficients are $\hat{\beta}_0 = 36.88$ and $\hat{\beta}_1 = 0.05151$. These are used to calculate the residuals $\hat{e}_i = y_i - (36.88 + 0.05151x_{i1})$. After ranking the residuals, we calculate $\text{SRWR}_{\text{reduced}} = \Sigma[\text{rank}(\hat{e}_i) - 25.5]\hat{e}_i = 31000$.

Next we use the residuals from the full model to obtain $\hat{\tau}$. From the 50 residuals we compute the 1275 ($= 50(51)/2$) pairwise averages. This includes the residuals themselves, since $A_{ii} = (\hat{e}_i + \hat{e}_i)/2 = \hat{e}_i$, as well as the averages $A_{ij} = (\hat{e}_i + \hat{e}_j)/2$ of distinct residuals for $i \neq j$. Put these pairwise averages in increasing order: $-81.803 < -67.586 < \cdots < 107.648 < 139.279$. Calculate $a = 50(51)/4 = 637.5$, $b = \sqrt{50(51)(101)/24} = 103.6$, $0.5 + 637.5 - (1.645)(103.6) = 467.6$, and $0.5 + 637.5 + (1.645)(103.6) = 808.4$. So $k_1 = 468$ and $k_2 = 808$. The 468th smallest pairwise average is -8.957 and the 808th smallest pairwise average is 9.343. The factor f is $\sqrt{50/46} = 1.043$. Now we obtain $\hat{\tau} = (1.043)\sqrt{50}\,[9.343 - (-8.957)]/[2(1.645)] = 41.01$.

The denominator of F_{NP} in formula (6.8) is $(p - q)c\hat{\tau} = (3 - 1) \times (51/\sqrt{48})(41.01) = 603.7$ and so we have $F_{NP} = (31000 - 27120)/603.7 = 6.421$. The degrees of freedom are $p - q = 2$ and $n - p - 1 = 46$. Looking in the F table in the Appendix, we see that the p-value is between 0.001 and 0.01. So one or both of the two explanatory variables X_2, percentage of young residents, and X_3, percentage of urban residents, contain significant information about the education expenditure of a state beyond the information contained in the variable X_1, personal income.

Simple Regression. Testing $\beta = 0$ in simple regression is the special case of testing $\beta_{q+1} = \cdots = \beta_p = 0$ in which $p = 1$ and $q = 0$. So we have two test statistics we could use, $|t_{NP}|$ in (6.5) or F_{NP} in (6.8). The p-value for $|t_{NP}|$ is calculated to be $\text{Prob}[|T| \geq |t_{NP}|]$, where T is a random variable having a t distribution with $n - 2$ degrees of freedom, or equivalently, $\text{Prob}[F \geq |t_{NP}|^2]$, where F is a random variable having an F distribution with 1 and $n - 2$ degrees of freedom. The p-value for F_{NP} is calculated to be $\text{Prob}[F \geq F_{NP}]$, where F is a random variable having an F distribution with 1 and $n - 2$ degrees of freedom. This leads us to ask if perhaps $|t_{NP}|^2 = F_{NP}$, but the answer is no.

For the forearm length data, $|t_{NP}| = 3.973$, so $|t_{NP}|^2 = 15.78$, whereas $F_{NP} = 20.01$. For either 15.78 or 20.01, the F table in the Appendix shows that the p-value is less than 0.001, which indicates that $\beta \neq 0$. However, sometimes the two test statistics can yield contradictory conclusions. If we test $\beta = 0$ for the birth rate data in Table 1.3, the value of $|t_{NP}|^2$ is 4.346, which gives a p-value greater than 0.05, whereas the value of F_{NP} is 8.616, which gives a p-value close to 0.01. It is common practice to accept the possibility that $\beta = 0$ when the p-value is greater than 0.05 and to conclude that $\beta \neq 0$ when the p-value is smaller than 0.05.

6.8 COMPUTATION

The nonparametric estimates minimizing (6.6) and the nonparametric test statistic (6.8) can be obtained from the statistical package Minitab by using its RANK REGRESSION command.

You could write your own computer program by following the descriptions of the algorithms presented in Sections 6.6 and 6.7. To check such a program, it could be applied to the test case data in Table 3.2. For simple regression using $X = X_1$, the estimated regression line is $\hat{Y} = 30.00 + 1.750X$ and the test statistic for testing $\beta = 0$ is $|t| = 1.965$. For multiple regression using both X_1 and X_2, the estimated regression equation is $\hat{Y} = 38.64 + 1.286X_1 - 0.2857X_2$ and the test statistic for testing $\beta_1 = \beta_2 = 0$ is $F_{NP} = 3.436$. The estimate $\hat{\tau}$ is 4.362.

NOTES

6.1a. The terms "parametric" and "nonparametric" are not precise. When Legendre originally proposed the least-squares method, the justification he gave was simply that it minimized the sum of squares of the deviations, with no reference to the distribution of the data. From this viewpoint, least-squares estimation is nonparametric. But least-squares estimates cannot be guaranteed to perform well unless the distribution of the error population is normal.

6.1b. We are using the term "nonparametric regression" to refer to regression procedures based on ranks. The term is also used to refer to procedures for constructing smooth estimates of nonlinear regression functions, such as in Härdle (1990).

6.1c. The plant heights that Galton analyzed were collected by Charles Darwin in an experiment to compare self-fertilization and cross-fertilization in plants. To see which mode of fertilization tends to produce taller plants, he measured the heights of 15 self-fertilized plants and 15 cross-fertilized plants of the same variety grown under the same conditions. Galton first arranged the heights of the selfed plants in increasing order and did the same for the crossed plants. He then compared the ith tallest selfed plant with the ith tallest crossed plant. In 13 out of the 15 comparisons, the crossed plant was taller.

6.3a. A *weighted average* of a list of numbers $x_1, \ldots, x_n$ is a linear combination $w_1x_1 + \cdots + w_nx_n$ in which the coefficients w_i are nonnegative and add to 1. The coefficient w_i, called the *weight* of x_i, is the proportion of influence that x_i has on the value of the weighted average. The ordinary average $\bar{x}$ is the special case in which the weights are all equal, that is, $w_i = 1/n$.

If $c_1, \ldots, c_n$ are nonnegative numbers, not all 0, then a weighted average of the x_i's with weights proportional to the c_i's is obtained by setting $w_i = c_i/\Sigma c_i$. The least-squares estimate $\hat{\beta}_{\text{LS}}$ is a weighted average of the pairwise slopes b_{ij} with each weight proportional to the squared x-distance $(x_i - x_j)^2$ between the two data points determining the slope. Note that the undefined pairwise slopes with $x_i = x_j$ are automatically ignored since they are given weight 0. As mentioned in Chapter 3, $\hat{\beta}_{\text{LS}}$ can also be expressed as a weighted average of the slopes $b_i = (y_i - \bar{y})/(x_i - \bar{x})$ of the lines between the data points (x_i, y_i) and the average point $(\bar{x}, \bar{y})$, with each weight proportional to the squared x-distance $(x_i - \bar{x})^2$.

6.3b. Estimation of the slope of the regression line by the ordinary median of the pairwise slopes was proposed by Thiel (1950). The weighted median of the pairwise slopes presented in Section 6.3 was proposed by Jaeckel (1972). The weighted median is a more efficient estimate than the ordinary median, at least for large sample sizes.

6.3c. The description of the ordinary median is not quite correct if the length of the list of numbers is an even integer. In this case, when the numbers are put in increasing order, there are two numbers, rather than just one, that are in the middle of the ordered list. The average of the two middle numbers is the median.

6.3d. Consider the weighted median of $x_1, \ldots, x_n$ with weights $w_i = 1/n$ for all i. First we put the numbers x_i in increasing order. By changing the indices, we can arrange so that $x_1 \le x_2 \le \cdots \le x_n$. We want to find an index k satisfying conditions (6.1), $w_1 + w_2 + \cdots + w_{k-1} < 0.5$ and $w_1 +$

$w_2 + \cdots + w_{k-1} + w_k > 0.5$, that is, satisfying $(k-1)(1/n) < 0.5$ and $k(1/n) > 0.5$, that is, $\frac{1}{2}n < k < \frac{1}{2}n + 1$. If n is an odd integer, then $\frac{1}{2}n + \frac{1}{2}$ is an integer and so we can set $k = \frac{1}{2}n + \frac{1}{2} = \frac{1}{2}(n+1)$. The weighted median is defined to be x_k, which is simply the ordinary median. If n is an even integer, then the index $k = \frac{1}{2}n$ satisfies $w_1 + w_2 + \cdots + w_k = k(1/n) = 0.5$. This situation is discussed in parentheses following (6.1). The weighted median is defined to be $(x_k + x_{k+1})/2$, which again is simply the ordinary median.

6.3e. The differences $y_i - \beta x_i$ are centered around α in the sense that they can be regarded as a random sample drawn from a population having a mean of α. In simple linear regression, the model is $y_i = \alpha + \beta x_i + e_i$, where the quantities e_i are a random sample drawn from a population having a mean of 0. Hence $y_i - \beta x_i = \alpha + e_i$, and the quantities $\alpha + e_i$ can be regarded as a random sample drawn from a population having a mean of α.

It would be natural to estimate the mean α of this population by the mean of the estimated differences $y_i - \hat{\beta} x_i$, but the mean is thought to be too sensitive to outliers, so the median of the estimated differences is used instead. If the population of errors has a symmetric distribution, then α is not only the mean but also the median of the population from which the differences $y_i - \beta x_i$ are drawn, and so it is natural to estimate α by the median of the estimated differences. If we are not comfortable in assuming that the population of errors has a symmetric distribution, then we might prefer to modify the model by assuming that the population of errors has a median of 0 rather than a mean of 0.

6.3f. A *weighted sum* of a list of numbers $x_1, \ldots, x_n$ is a linear combination $w_1 x_1 + \cdots + w_n x_n$ in which the coefficients w_i are nonnegative.

6.3g. The estimation of regression coefficients by minimizing (6.2) was proposed by Jaeckel (1972). In simple regression, the same estimate of β was proposed earlier, in a different form, by Adichie (1967). He derived the estimate by using the statistic U in (6.5). Define $U(b)$ to be U with y_i replaced by $y_i - bx_i$. The nonparametric estimate $\hat{\beta}$ can be obtained as a value of b for which $U(b)$ is closest to 0. To see that Adichie's estimate coincides with Jaeckel's estimate, note that the derivative of (6.3) with respect to b is $-U(b)$.

6.3h. Another justification for why it is sensible to minimize (6.2) is provided by the fact that (6.2) can be expressed as $\frac{1}{2}$ times the sum of the absolute values of the differences between all pairs of residuals (see Hettmansperger and Aubuchon, 1988). (This is proved in Note 6.3l; substitute $\hat{e}_i$ for x_i.) Therefore, choosing a value of b to minimize (6.2) causes the residuals $y_i - bx_i$ to have values that are relatively close to one another. By then choosing an appropriate value of a, the residuals $y_i - (a + bx_i)$ should be relatively close to 0.

6.3i. In Note 6.3h it is remarked that minimizing (6.2) is equivalent to minimizing the dispersion of the residuals as measured by $\Sigma|\hat{e}_i - \hat{e}_j|$, where the sum is over all pairs of residuals. More generally, a reasonable measure of dispersion is given by (*) $\Sigma c(r_i, r_j)|\hat{e}_i - \hat{e}_j|$ for nonnegative coefficients $c(r, s)$ depending on the ranks r_i of the residuals. Besides nonnegativity, let us require the coefficients to satisfy $c(s, r) = c(r, s)$ (because $|\hat{e}_j - \hat{e}_i| = |\hat{e}_i - \hat{e}_j|$), $c(r, r) = 0$ (because $|\hat{e}_i - \hat{e}_i| = 0$), and $c(r, s) \leq c(r, t)$ for $r < s < t$ (to give larger differences between residuals at least as much weight as smaller differences). It can be shown that minimizing (*) is equivalent to minimizing $\Sigma a(r_i)\hat{e}_i$ with coefficients $a(r)$ that are nondecreasing for $r = 1, \ldots, n$ and sum to 0. The $a(r)$ are called *scores*. Different scores can be chosen to obtain different estimates of β. The two most common choices of scores are the Wilcoxon scores, $a(r) = r - \frac{1}{2}(n + 1)$, which are used in (6.2), and the normal scores, $a(r) = \Phi^{-1}(r/(n + 1))$, where Φ denotes the cumulative distribution function of the standard normal distribution.

6.3j. Estimates obtained by minimizing the sum $\Sigma\, \text{rank}(|\hat{e}_i|)|\hat{e}_i|$ are called signed-rank estimates to distinguish them from the rank estimates obtained by minimizing sum (6.2). When the error distribution is symmetric and the sample size is large, the two kinds of estimates behave approximately the same (see Hettmansperger and McKean, 1983, Section 1). An algebraic relationship between the two sums is given by Hettmansperger and Aubuchon (1988).

6.3k. To see that the graph of function (6.3) consists of a series of line segments joined together at the points $b = b_{ij}$, consider the value of $\text{rank}(y_i - bx_i)$. It is equal to the number of data points (x_j, y_j) for which $y_j - bx_j \leq y_i - bx_i$. By rearranging this inequality, we see that $\text{rank}(y_i - bx_i)$ equals the number of data points for which $x_j < x_i$ and $b \geq (y_i - y_j)/(x_i - x_j) = b_{ij}$ plus the number of data points for which $x_j > x_i$ and $b \geq (y_i - y_j)/(x_i - x_j) = b_{ij}$ plus the number of data points for which $x_j = x_i$ and $y_j \leq y_i$. The important thing to notice about this description of $\text{rank}(y_i - bx_i)$ is that it depends on b only through the position of b relative to the points b_{ij}. This implies that all the ranks involved in (6.3) are constant as b varies within any one of the intervals between the b_{ij}. Therefore, in each such interval, (6.3) can be written as the linear function $c_0 + c_1 b$, where $c_0 = \Sigma[r_i - \frac{1}{2}(n + 1)]y_i$, $c_1 = -\Sigma[r_i - \frac{1}{2}(n + 1)]x_i$, and r_i is the constant rank of $y_i - bx_i$ in that interval.

6.3l. Let us verify that, as b varies from $-\infty$ to $+\infty$, the slope of function (6.3) goes from $-\frac{1}{2}T$ to $\frac{1}{2}T$, where $T = \Sigma|x_i - x_j|$. (The summation in the definition of T is taken over the $n(n - 1)/2$ sets of two distinct integers i and j between 1 and n.) We have seen that the slope of (6.3) is $-\Sigma[\text{rank}(y_i - bx_i) - \frac{1}{2}(n+1)]x_i$. For very large negative values of b, $\text{rank}(y_i - bx_i) = \text{rank}(x_i)$, so the slope is $-\Sigma[\text{rank}(x_i) - \frac{1}{2}(n+1)]x_i$. For very large positive values of b, $\text{rank}(y_i - bx_i) = \text{rank}(-x_i) = (n+1) - \text{rank}(x_i)$, so the slope is $-\Sigma[(n+1) - \text{rank}(x_i) - \frac{1}{2}(n+1)]x_i = \Sigma[\text{rank}(x_i) - \frac{1}{2}(n+1)]x_i$. So we want to verify the equation (*) $\Sigma[\text{rank}(x_i) - \frac{1}{2}(n+1)]x_i = \frac{1}{2}\Sigma|x_i - x_j|$.

Since $|x_i - x_j| = x_i - x_j$ if $x_i > x_j$ and $|x_i - x_j| = x_j - x_i$ if $x_i < x_j$, it follows that $\Sigma|x_i - x_j| = \Sigma(L_i - G_i)x_i$, where L_i = the number of x_j's less than x_i and G_i = the number of x_j's greater than x_i. It is convenient to assume that the x_i are distinct, so that $\text{rank}(x_i) = L_i + 1$ and $L_i + G_i = n - 1$. (If some of the x_i's are equal and midranks are used, equation (*) is still true.) Hence $L_i - G_i = L_i - (n - 1 - L_i) = 2L_i - n + 1 = 2[\text{rank}(x_i) - 1] - n + 1 = 2[\text{rank}(x_i) - \frac{1}{2}(n+1)]$. Therefore $\Sigma|x_i - x_j| = 2\Sigma[\text{rank}(x_i) - \frac{1}{2}(n+1)]x_i$.

6.4a. For more about test statistic (6.5), see Hettmansperger (1984, pp. 224–226). Perhaps the earliest rank-based nonparametric test of the slope in simple linear regression was given by Terry (1952). His test is what is sometimes called a linear rank test with normal scores. The test in Section 6.4 is a linear rank test with Wilcoxon scores. See Hájek and Šidák (1967, Section III.3.1).

6.4b. Why does the standard deviation of $\hat{\beta}_{LS}$ have to be estimated whereas the standard deviation of U is known? The randomness of the random variable $\hat{\beta}_{LS}$ comes from the randomness of the y_i's. As seen in Note 3.4a, the SD of $\hat{\beta}_{LS}$ depends on the SD of the y_i's, which is equal to the SD of the errors e_i, which is σ. The parameter σ is unknown and must be estimated. The randomness of the random variable U also comes from the randomness of the y_i's but the dependence of U on y_i is only through $\text{rank}(y_i)$. When the null hypothesis $\beta = 0$ is true, the y_i's can be regarded as a random sample from a single population, and so their ranks are simply a random permutation of the integers 1 through n. Therefore the distribution of the ranks of the y_i's does not involve any unknown parameters, and hence neither does the distribution of U.

SD(U) is the correct standard deviation of U only if the null hypothesis is assumed to be true. But est.SD($\hat{\beta}_{LS}$) is a valid estimate of the standard deviation of $\hat{\beta}_{LS}$ regardless of whether the null hypothesis is true.

6.4c. To be conservative, the t distribution could be used to calculate the p-value for test statistic (6.5) instead of the standard normal distribution. For example, the test statistic for the forearm length data is $|t| = 3.973$. By using the t distribution with 31 degrees of freedom, the p-value is 0.00038; whereas by using the standard normal distribution, the p-value is 0.00007. In either case, the p-value is very small and we conclude that there is a significant relationship between forearm length and height, but note that the p-value using the t distribution is larger, which is in the conservative direction.

6.4d. The ranks $r_i = \text{rank}(y_i)$ are simply a permutation of the integers 1 through n. So the average rank is the average of the integers 1 through n; that is, $\bar{r} = (1 + 2 + \cdots + n)/n = \frac{1}{2}(n + 1)$. If we regard the integers 1 through n as a sample, the sample variance is

$$\begin{aligned}
&\frac{1}{n-1} \sum \left[i - \tfrac{1}{2}(n+1)\right]^2 \\
&\quad = \frac{1}{n-1}\left\{\sum i^2 - (n+1)\sum i + n\left(\tfrac{1}{4}\right)(n+1)^2\right\} \\
&\quad = \frac{1}{n-1}\left\{\tfrac{1}{6}n(n+1)(2n+1) - (n+1)\tfrac{1}{2}n(n+1) + \tfrac{1}{4}n(n+1)^2\right\} \\
&\quad = \tfrac{1}{12}n(n+1).
\end{aligned}$$

6.6a. The estimation of regression coefficients by minimizing (6.6) was proposed by Jaeckel (1972). Jaeckel's estimates are essentially the same as those of Jurečková (1971); they coincide for infinitely large samples. One approach to minimizing function (6.6) would be to take its partial derivatives with respect to the b_j's, equate these derivatives to 0, and try to solve for the b_j's. In general there may not be a solution that makes all the derivatives exactly 0, and so Jurečková chose the b_j's to minimize the sum of the absolute values of the derivatives.

Jaeckel and Jurečková considered arbitrary scores (see Note 6.3i), not just the Wilcoxon scores that we use in this chapter. An estimate of $\boldsymbol{\beta}$ can be obtained by minimizing the function $g(\boldsymbol{b}) = \Sigma a(r_i)\hat{e}_i$, where $\hat{e}_i = y_i - \boldsymbol{b}'\boldsymbol{x}_i$, r_i is the rank of $\hat{e}_i$, and $a(r_i)$ is the score. Function (6.6) (or, in vector notation, (6.7)) is the special case in which $a(r) = r - \frac{1}{2}(n + 1)$.

6.6b. The function $g(\boldsymbol{b})$ is known to be well behaved with respect to minimization. It is a convex function (see Hettmansperger, 1984, p. 234), which implies that the condition $\nabla g(\boldsymbol{b}^0) = \boldsymbol{0}$ is sufficient to ensure that $\boldsymbol{b}^0$ minimizes the function. However, there does not necessarily exist a point $\boldsymbol{b}^0$ satisfying $\nabla g(\boldsymbol{b}^0) = 0$, that is, at which all the partial derivatives exist and are 0. A necessary and sufficient criterion for $\boldsymbol{b}^0$ to minimize $g(\boldsymbol{b})$ is that at $\boldsymbol{b}^0$, for each $j = 1, \ldots, p$, the left-hand partial derivative with respect to b_j is nonpositive and the right-hand partial derivative is nonnegative.

6.6c. The derivative of $g(\boldsymbol{b}^0 + t\boldsymbol{d})$ with respect to t, according to the chain rule, is $(\partial/\partial t)g(\boldsymbol{b}^0 + t\boldsymbol{d}) = (\nabla g(\boldsymbol{b}^0 + t\boldsymbol{d}))'\boldsymbol{d}$. At $t = 0$ it is $(\nabla g(\boldsymbol{b}^0))'\boldsymbol{d}$. To simplify notation, let $\boldsymbol{h} = \nabla g(\boldsymbol{b}^0)$. We want to choose $\boldsymbol{d}$ so that $\boldsymbol{h}'\boldsymbol{d} < 0$. Since $\boldsymbol{h}'\boldsymbol{h} = \Sigma h_i^2 > 0$ (assuming $\boldsymbol{h} \neq \boldsymbol{0}$), we can choose $\boldsymbol{d} = -\boldsymbol{h}$. Also,

$h'(X_c'X_c)^{-1}h > 0$ (because $h'(X_c'X_c)^{-1}h = f'f$, where $f = X_c(X_c'X_c)^{-1}h$), and so we can choose $d = -(X_c'X_c)^{-1}h = -(X_c'X_c)^{-1}\nabla g(b^0)$.

6.6d. The direction vector $d = -(X_c'X_c)^{-1}\nabla g(b^0)$ was suggested by McKean and Hettmansperger (1978, p. 574). The matrix $(X_c'X_c)^{-1}$ is approximately proportional to the variance–covariance matrix of the nonparametric regression estimates of $\beta_1, \ldots, \beta_p$. See the subsection on the parameter τ in Section 6.7 and Note 6.7e.

6.6e. The accuracy of estimates obtained from an iterative algorithm is discussed in Notes 5.3g, 5.3h, and 5.6c. For the education expenditure data, we iterated the nonparametric estimation algorithm until two successive vectors of estimates satisfied the condition that the relative differences were less than 10^{-8}. However, the function (6.7), which we are trying to minimize, appears to be rather "flat" and so we are only confident in the accuracy of the estimates to about three significant digits. Using the least-squares estimates (0.07239, 15.52, −0.04269) as initial estimates, the algorithm converged to (0.05868, 12.54, 0.3372). Using the slightly different initial estimates (0.07, 16.0, −0.04), the algorithm converged to (0.05857, 12.56, 0.3380).

6.6f. Rather than require convergence of the estimation algorithm, it is sometimes sufficient to do only one or two iterations of the algorithm. McKean and Hettmansperger (1978) found that, provided the sample size is large and the error distribution does not have extremely heavy tails, the estimates obtained from only one iteration are generally quite close to the estimates that would be obtained by iterating until convergence. This can save considerable computation time for large data sets.

6.6g. When ties occur among the differences $y_i - b'x_i$, we need to specify how to assign their ranks. We used midranks in our calculations for the education expenditure data. But if there are not many ties, it should not make much difference how ties are handled. To avoid calculation of midranks, one could break ties in some arbitrary manner such as according to the order of their indices.

6.7a. Test statistic F_{NP} in (6.8) was introduced by McKean and Hettmansperger (1976). Also see Hettmansperger (1984, Subsection 5.3.1). To see that our formula for F_{NP} agrees with Hettmansperger's book, note that his sum of rank-weighted residuals in (5.2.5) using $a(i)$ from (5.2.11) is equal to our SRWR multiplied by $\sqrt{12}/(n+1)$. If we had used his definition of SRWR, we would have defined $c = \frac{1}{2}$ in (6.8).

In Chapter 6 we estimate β by minimizing $g(b) = \Sigma a(r_i)\hat{e}_i$ with $a(r) = r - \frac{1}{2}(n+1)$; whereas Hettmansperger (1984) and others use $a(r) = [\sqrt{12}/(n+1)][r - \frac{1}{2}(n+1)]$. Of course the constant factor $\sqrt{12}/(n+1)$ does not affect the minimization. The factor serves simply to standardize the scores $a(r)$ so that the average of the squared scores is approximately 1. (The average of the squared scores is exactly equal to $(n-1)/(n+1)$; see Note 6.4d.)

6.7b. Two other tests are available for testing the coefficients in a multiple linear regression model: the aligned rank test, proposed by Koul (1970) and developed by Sen and Puri (1977) and Adichie (1978), and the Wald test, proposed by Hettmansperger and McKean (1983).

The aligned rank test can be described in terms of the function $g(\boldsymbol{b})$ in (6.7). Recall that the estimation procedure in Section 6.6 is based on the fact that the minimum of $g(\boldsymbol{b})$ should occur for a vector that is near the parameter vector $\boldsymbol{\beta}$. This implies that the partial derivatives of $g(\boldsymbol{b})$ should be close to 0 at $\boldsymbol{b} = \boldsymbol{\beta}$; that is, $(\partial g/\partial b_j)(\boldsymbol{\beta}) \approx 0$. To test the hypothesis $\beta_{q+1} = \cdots = \beta_p = 0$, we first estimate $\boldsymbol{\beta}$ under the assumption that the hypothesis is true, by applying the procedure in Section 6.6 to the reduced model to obtain $\hat{\beta}_1, \ldots, \hat{\beta}_q$ and setting $\hat{\beta}_{q+1} = \cdots = \hat{\beta}_p = 0$. Let $\hat{\boldsymbol{\beta}}_0$ denote this vector of estimates. If the hypothesis is true, then the vector $\hat{\boldsymbol{S}}_2$ with entries $(\partial g/\partial b_j)(\hat{\boldsymbol{\beta}}_0)$, $j = q+1, \ldots, p$, should be near $\mathbf{0}$. The aligned rank test statistic is $A = \hat{\boldsymbol{S}}_2' \boldsymbol{V}^{-1} \hat{\boldsymbol{S}}_2$, where $\boldsymbol{V} = \boldsymbol{X}_{c2}'\boldsymbol{X}_{c2} - \boldsymbol{X}_{c2}'\boldsymbol{X}_{c1}(\boldsymbol{X}_{c1}'\boldsymbol{X}_{c1})^{-1}\boldsymbol{X}_{c1}'\boldsymbol{X}_{c2}$ and $\boldsymbol{X}_{c1}$ and $\boldsymbol{X}_{c2}$ are, respectively, columns 1 through q and columns $q+1$ through p of the matrix $\boldsymbol{X}_c$ of centered explanatory variables. The hypothesis is rejected if A is too large. The p-value is calculated from the chi-squared distribution with $p - q$ degrees of freedom. See Hettmansperger (1984, Section 5.3.2). The phrase "aligned rank" comes from the fact that the test involves the ranks of the residuals $y_i - \hat{\boldsymbol{\beta}}_0'\boldsymbol{x}_i$, which can be regarded as alignments of the observations y_i. An aligned rank test uses ranks of residuals, whereas a "pure" rank test uses the ranks of the observations.

The Wald test of the hypothesis $\beta_{q+1} = \cdots = \beta_p = 0$ is based directly on the vector of estimates $\hat{\boldsymbol{\beta}}_2 = (\hat{\beta}_{q+1}, \ldots, \hat{\beta}_p)$ from the full model. If the hypothesis is true, then $\hat{\boldsymbol{\beta}}_2$ should be near $\mathbf{0}$. The Wald test statistic is $W = \hat{\boldsymbol{\beta}}_2'\boldsymbol{V}\hat{\boldsymbol{\beta}}_2/(p-q)\hat{\tau}^2$, where $\boldsymbol{V}$ is from the preceding paragraph and $\hat{\tau}$ is obtained from (6.9). The hypothesis is rejected if W is too large. The p-value is calculated from the F distribution with $p - q$ and $n - p - 1$ degrees of freedom. See Hettmansperger (1984, Section 5.3.3).

All three tests are equivalent for infinitely large samples. In simulation studies with finite samples, Hettmansperger and McKean (1983) found the test in Section 6.7 to be more stable in its validity and power (see Note 3.5b). The aligned rank test avoids the problem of estimating τ.

6.7c. Formula (6.9) for estimating τ is recommended by Hettmansperger and McKean (1983, Section 4). The factor f is *ad hoc* and has no theoretical justification. In simulated examples, f was found to improve the accuracy of the p-value. See Draper (1988, Section 4) for a review of results on the estimation of τ.

6.7d. Test statistic F_{NP}, regarded as a random variable, has approximately the same distribution as $G/(p-q)$, where G has a χ^2 distribution with $p-q$ degrees of freedom, provided the null hypothesis $\beta_{q+1} = \cdots = \beta_p = 0$ is true and the sample size n is large. (See Hettmansperger, 1984, Theorem 5.3.1.) To be more mathematically precise, for any fixed number c, the limit of $\text{Prob}[F_{\text{NP}} \geq c]$ approaches $\text{Prob}[G/(p-q) \geq c]$ as n becomes very large.

There is no theoretically justified connection between F_{NP} and the F distribution, but nevertheless in a number of examples having smaller sample sizes, statistical researchers have found that a more accurate approximation of the p-value is obtained if the F distribution with $p-q$ and $n-p-1$ degrees of freedom, rather than $G/(p-q)$, is used to approximate the distribution of F_{NP} (see Hettmansperger, 1984, p. 266).

6.7e. The matrix $\boldsymbol{W}$ is defined to be the $p \times p$ matrix obtained from $(\boldsymbol{X}'\boldsymbol{X})^{-1}$ by omitting the first row and first column. Another way to obtain $\boldsymbol{W}$ is as $\boldsymbol{W} = (\boldsymbol{X}_c'\boldsymbol{X}_c)^{-1}$, where $\boldsymbol{X}_c$ is the matrix of centered explanatory variables. The value x_{ij} of the jth explanatory variable for the ith unit is centered by subtracting the average $\bar{x}_j$ of the jth explanatory variable for all n units.

Express $\boldsymbol{X}$ in partitioned form as $\boldsymbol{X} = (\mathbf{1}, \mathbf{Z})$, where $\mathbf{1}$ is a column of 1's and $\mathbf{Z}$ is the matrix of explanatory variables. Then

$$\boldsymbol{X}'\boldsymbol{X} = \begin{bmatrix} \mathbf{1}'\mathbf{1} & \mathbf{1}'\mathbf{Z} \\ \mathbf{Z}'\mathbf{1} & \mathbf{Z}'\mathbf{Z} \end{bmatrix} = \begin{bmatrix} n & \mathbf{1}'\mathbf{Z} \\ \mathbf{Z}'\mathbf{1} & \mathbf{Z}'\mathbf{Z} \end{bmatrix}$$

In the case of simple regression, $\mathbf{Z}$ is a vector and $\boldsymbol{X}'\boldsymbol{X}$ is a 2×2 matrix, which is easy to invert. The entry that is left after omitting the first row and column of $(\boldsymbol{X}'\boldsymbol{X})^{-1}$ is $n/[n(\mathbf{Z}'\mathbf{Z}) - (\mathbf{Z}'\mathbf{1})(\mathbf{1}'\mathbf{Z})] = [\mathbf{Z}'\mathbf{Z} - (1/n)\mathbf{Z}'\mathbf{1}\mathbf{1}'\mathbf{Z}]^{-1}$. It can be shown that this formula also holds for multiple regression, that is, $\boldsymbol{W} = [\mathbf{Z}'\mathbf{Z} - (1/n)\mathbf{Z}'\mathbf{1}\mathbf{1}'\mathbf{Z}]^{-1}$.

The matrix $\boldsymbol{X}_c$ of centered variables is obtained by subtracting from each entry of $\mathbf{Z}$ the average of the entries in that column. In matrix terms, $\boldsymbol{X}_c = \mathbf{Z} - (1/n)\mathbf{1}\mathbf{1}'\mathbf{Z}$. (The row vector $\mathbf{1}'\mathbf{Z}$ contains the p sums of the columns of $\mathbf{Z}$, the row vector $(1/n)\mathbf{1}'\mathbf{Z}$ contains the p averages of the columns of $\mathbf{Z}$, and $(1/n)\mathbf{1}\mathbf{1}'\mathbf{Z}$ repeats the row of averages n times.) A little matrix algebra shows that $\boldsymbol{X}_c'\boldsymbol{X}_c = \mathbf{Z}'\mathbf{Z} - (1/n)\mathbf{Z}'\mathbf{1}\mathbf{1}'\mathbf{Z}$.

6.7f. We have defined two estimates of β_0. Let $d_i = y_i - (\hat{\beta}_1 x_{i1} + \cdots + \hat{\beta}_p x_{ip})$ and define $\hat{\beta}_0^{(d)}$ to be the median of the d_i's and $\hat{\beta}_0^{(dd)}$ to be the median of the pairwise averages $(d_i + d_j)/2$. The second estimate is appropriate only if the error distribution is approximately symmetric. If symmetry holds, then, at least when the sample size is large, $\hat{\beta}_0^{(dd)}$ has a smaller SD than $\hat{\beta}_0^{(d)}$ if and only if the parameter τ satisfies $\tau < 1/(2\theta)$, where θ is the probability density of the error distribution at 0. (See Hettmansperger, 1984, pp. 250–251.) The quantity $1/(2\theta)$ appears in LAD regression in a role similar to that of τ in nonparametric regression. (In Chapter 4 the symbol τ is used to denote the quantity $1/(2\theta)$.)

6.7g. Let us show that $\tau/\sigma = 1.023$ when the error population has a normal distribution. If the distribution of a randomly selected error is normal with standard deviation σ, then the distribution of the difference of two randomly selected errors is normal with standard deviation $\sqrt{2}\,\sigma$. Hence the probability density function for the error differences is

$$f(t) = \frac{1}{\sqrt{4\pi\sigma^2}} \exp\left(-\frac{t^2}{4\sigma^2}\right)$$

Now $\gamma = f(0) = 1/\sqrt{4\pi\sigma^2}$ and $\tau = 1/(\sqrt{12}\,\gamma) = \sigma\sqrt{\pi/3}$, so $\tau/\sigma = \sqrt{\pi/3} = 1.023$.

6.7h. Let $p = \text{Prob}[A_{(k_1)} < \mu < A_{(k_2)}]$. We can show that $p \approx 0.90$. Note that the event $A_{(k)} < \mu$ is equivalent to the event that at least k of the pairwise averages A_i are less than μ. Therefore, letting x be the number of pairwise averages that are less than μ, we can write $p = \text{Prob}[k_1 \le x < k_2]$. To adjust for the discreteness of x, we write this probability as $\text{Prob}[k_1 - \frac{1}{2} < x < k_2 - \frac{1}{2}]$.

Now we use the fact (see Hettmansperger, 1984, p. 39) that x has a distribution that, at least for large n, is approximately normal with mean a and standard deviation b. Hence $0.90 \approx \text{Prob}[a - (1.645)b < x < a + (1.645)b] = \text{Prob}[k_1 - \frac{1}{2} < x < k_2 - \frac{1}{2}] = p$.

Note that the two confidence intervals that we have compared in order to justify formula (6.9) are both based on normal approximations. To construct the interval from $\hat{\mu} - (1.645)\hat{\tau}/\sqrt{n}$ to $\hat{\mu} + (1.645)\hat{\tau}/\sqrt{n}$, we approximated the distribution of $\hat{\mu}$ by a normal distribution. To construct the interval from $A_{(k_1)}$ to $A_{(k_2)}$, we approximated the distribution of x by a normal distribution.

Additional Reading. To read more about nonparametric regression, see the books by Hettmansperger (1984, Chapter 5) and Puri and Sen (1985, Part 2). The first book is a little more oriented toward application. For a review of nonparametric regression methods, see the articles by Aubuchon and Hettmansperger (1984) and Draper (1988).

REFERENCES

Adichie, J. N. (1967). Estimates of regression parameters based on rank tests. *Annals of Mathematical Statistics*, vol. 38, pp. 894–904.

Adichie, J. N. (1978). Rank tests of sub-hypotheses in the general linear regression. *Annals of Statistics*, vol. 6, pp. 1012–1026.

Aubuchon, J. C. (1982). *Rank Tests in the Linear Model: Asymmetric Errors*. Ph.D. thesis, Department of Statistics, Pennsylvania State University.

Aubuchon, J. C., and T. P. Hettmansperger (1984). On the use of rank tests and estimates in the linear model. In: P. R. Krishnaiah and P. K. Sen (eds.), *Handbook of Statistics: Nonparametric Methods*, vol. 4. Elsevier, Amsterdam.

Chatterjee, S., and B. Price (1977). *Regression Analysis by Example*. Wiley, New York.

Darwin, C. (1876). *The Effects of Cross- and Self-Fertilization in the Vegetable Kingdom*. John Murray, London.

Draper, D. (1988). Rank-based robust analysis of linear models. I. Exposition and review. *Statistical Science*, vol. 3, pp. 239–271.

Gunst, R. F., and R. L. Mason (1980). *Regression Analysis and Its Applications*. Marcel Dekker, New York.

Hájek, J., and Z. Šidák (1967). *Theory of Rank Tests*. Academic Press, New York.

Härdle, W. (1990). *Applied Nonparametric Regression*. Cambridge University Press, Cambridge.

Hettmansperger, T. P. (1984). *Statistical Inference Based on Ranks*. Wiley, New York.

Hettmansperger, T. P., and J. C. Aubuchon (1988). Comment in Draper (1988).

Hettmansperger, T. P., and J. W. McKean (1977). A robust alternative based on ranks to least squares in analyzing linear models. *Technometrics*, vol. 19, pp. 275–284.

Hettmansperger, T. P., and J. W. McKean (1983). A geometric interpretation of inferences based on ranks in the linear model. *Journal of the American Statistical Association*, vol. 78, pp. 885–893.

Jaeckel, L. A. (1972). Estimating regression coefficients by minimizing the dispersion of the residuals. *Annals of Mathematical Statistics*, vol. 43, pp. 1449–1458.

Jurečková, J. (1971). Nonparametric estimate of regression coefficients. *Annals of Mathematical Statistics*, vol. 42, pp. 1328–1338.

Koul, H. L. (1970). A class of ADF tests for subhypotheses in the multiple linear regression. *Annals of Mathematical Statistics*, vol. 41, pp. 1273–1281.

McKean, J. W., and T. P. Hettmansperger (1976). Tests of hypotheses based on ranks in the general linear model. *Communication in Statistics A*, vol. 5, pp. 693–709.

McKean, J. W., and T. P. Hettmansperger (1978). A robust analysis of the general linear model based on one-step *R*-estimates. *Biometrika*, vol. 65, pp. 571–579.

Puri, M. L., and P. K. Sen (1985). *Nonparametric Methods in General Linear Models*. Wiley, New York.

SAS Institute Inc. (1988). *SAS Procedures Guide*, release 6.03 ed. SAS Institute Inc., Cary, NC.

Sen, P. K., and M. L. Puri (1977). Asymptotically distribution-free aligned rank order tests for composite hypotheses for general multivariate linear models. *Zeitschrift für Wahrscheinlichkeitstheorie und Verbandte Gebiete*, vol. 39, pp. 175–186.

Terry, M. E. (1952). Some rank order tests which are most powerful against specific parametric alternatives. *Annals of Mathematical Statistics*, vol. 23, pp. 346–366.

Thiel, H. (1950). A rank-invariant method of linear and polynomial regression analysis I, II and III. *Koninklijke Nederlandse Akademie van Wetenschappen. Proceedings*, ser. A, vol. 53, pp. 386–392, 521–525, 1397–1412.

CHAPTER 7

Bayesian Regression

7.1 INTRODUCTION

The Bayesian approach to statistical analysis is different from the usual "classical" approach. In the classical approach, the data are the only source of information explicitly taken into account in constructing an estimate or test. In the Bayesian approach, an estimate or test is produced by combining the current data with information from past experience.

The basic formula that is used to incorporate past knowledge into statistical analysis was discovered by Thomas Bayes around 1760. Perhaps the first application to regression problems was made by Harold Jeffreys in 1939.

7.2 THE BAYESIAN APPROACH

The Bayesian approach is a general scheme that can be applied to a wide variety of statistical problems. Before applying it to regression problems, let us outline its general features. In order to analyze a set of data, we usually postulate a statistical model for it. For notation, let $\boldsymbol{y}$ denote the vector containing the data and let $\boldsymbol{\theta}$ denote the vector of unknown parameters of the model. (In a simple regression problem, $\boldsymbol{y} = (y_1, y_2, \ldots, y_n)$ and $\boldsymbol{\theta} = (\alpha, \beta, \sigma)$.) We want to use the data $\boldsymbol{y}$ to estimate, or test hypotheses about, some of the parameters in $\boldsymbol{\theta}$.

The statistical model states that the data vector $\boldsymbol{y}$, regarded as a random variable, has a certain probability distribution for each fixed value of the parameter vector $\boldsymbol{\theta}$. (The simple normal linear regression model states that, for fixed values of α, β, and σ, the distribution of y_i is normal with mean $\alpha + \beta x_i$ and variance σ^2 and the y_i's are independent of one another.) The distribution of $\boldsymbol{y}$ for a fixed value of $\boldsymbol{\theta}$ is called the *conditional distribution* of $\boldsymbol{y}$ given $\boldsymbol{\theta}$. In the Bayesian approach, we must also specify a distribution for $\boldsymbol{\theta}$. Before looking at the data, we must assess what we know or believe about likely values for the parameters and translate this knowledge or belief into the form of a probability distribution for $\boldsymbol{\theta}$. This is called the *prior distribution* of $\boldsymbol{\theta}$. The next step is to combine the prior distribution of $\boldsymbol{\theta}$ and the conditional distribution of $\mathbf{y}$ given $\boldsymbol{\theta}$ to obtain the conditional distribution of $\boldsymbol{\theta}$ given $\mathbf{y}$. For the particular data vector $\mathbf{y}$ that is observed, this is called the *posterior distribution* of $\boldsymbol{\theta}$. In this way, prior information about the parameters is updated by current data to yield posterior information.

Bayes's formula provides a means of combining the distributions of $\boldsymbol{\theta}$ and of $\mathbf{y}$ given $\boldsymbol{\theta}$ to obtain the distribution of $\boldsymbol{\theta}$ given $\mathbf{y}$. Often these distributions are presented in terms of their probability density functions $f(\boldsymbol{\theta})$, $f(\mathbf{y}|\boldsymbol{\theta})$, and $f(\boldsymbol{\theta}|\mathbf{y})$. Bayes's formula says that

$$f(\boldsymbol{\theta}|\mathbf{y}) = Cf(\boldsymbol{\theta})f(\mathbf{y}|\boldsymbol{\theta}) \tag{7.1}$$

where C is a quantity that does not involve $\boldsymbol{\theta}$.

Bayesian estimates of, or tests of hypotheses about, the parameters in $\boldsymbol{\theta}$ are obtained from the posterior distribution of $\boldsymbol{\theta}$. For example, we can estimate $\boldsymbol{\theta}$ by the mean of its posterior distribution.

7.3 AN EXAMPLE OF SIMPLE REGRESSION

Due to weather patterns in the northwest corner of the United States, rainfall in Seattle is related to rainfall in Portland. To investigate this relationship in more detail, let us consider the data in Table 7.1. It lists the annual amounts of rainfall (in inches) in both cities for the years 1980 through 1990.

Table 7.1 Rainfall Data

Year	Rainfall in Seattle (Y)	Rainfall in Portland (X)
1980	35.60	42.41
1981	35.40	34.29
1982	39.32	43.04
1983	40.93	47.19
1984	36.99	37.50
1985	25.13	22.48
1986	38.34	35.04
1987	29.93	29.91
1988	32.98	31.72
1989	34.69	30.05
1990	44.75	32.86

Source: NOAA (1990).

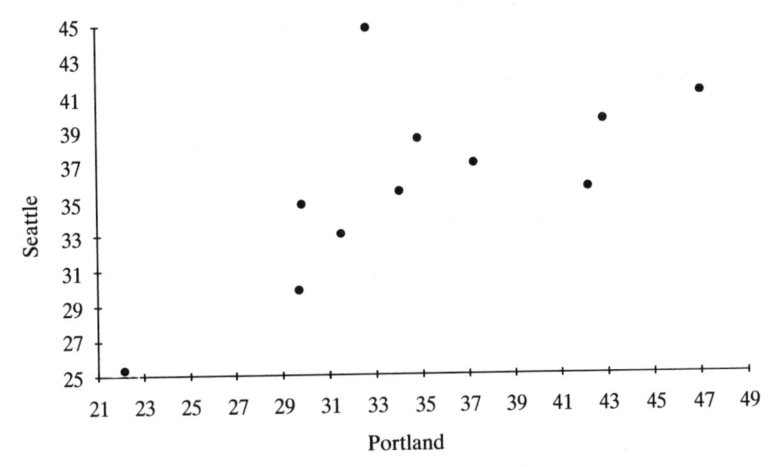

Figure 7.1 Plot of the rainfall data.

A plot of the 11 data points, with the coordinates x = rainfall in Portland and y = rainfall in Seattle, is shown in Figure 7.1. The overall pattern of the plot is roughly linear. By applying the simple linear regression model $Y = \alpha + \beta X + e$ to these data, we will be able to see how well the annual rainfall in Seattle can be expressed as a linear function of the annual rainfall in Portland.

7.4 ESTIMATING THE REGRESSION LINE

First we must specify a probability distribution for the data vector. The most widely used distribution is the normal distribution. Let us assume the normal linear regression model. Then, given the parameters α, β, and σ, the data vector $\mathbf{y} = (y_1, \ldots, y_n)$ has a multivariate normal distribution.

The Bayesian method also requires that we specify a joint prior distribution for the parameters α, β, and σ. Theoretically speaking, any distribution can be chosen as the prior distribution. But practically speaking, calculation is much easier if we choose a prior distribution that conveniently combines with the distribution of the data vector when we calculate the posterior distribution. Two types of prior distributions are presented below.

Bayesian estimation is not much simpler to justify for simple regression than for multiple regression, and so parts of the justification are deferred to Section 7.7.

Using a Noninformative Prior Distribution. The prior information or belief that a person has about the parameters depends on the person. Suppose we have no prior information or belief. That is, before looking at the data in Table 7.1, suppose we have no idea what the values of α, β, and σ might be, other than that σ must be positive. Here we need a "noninformative" prior distribution that expresses our ignorance. For this purpose it is common to use the density function

$$f(\alpha, \beta, \sigma) = 1/\sigma \tag{7.2}$$

This is essentially saying that, as far as we know before looking at the data, all values of α are equally likely, all values of β are equally likely, and all values of $\log \sigma$ are equally likely.

It can be argued that no prior distribution is completely noninformative. For instance, having no prior information about σ would seem to be the same as having no prior information about $\log \sigma$. Yet when we try to formulate these two seemingly equivalent statements of prior ignorance, we use the density function $f(\sigma) = 1$ to formulate the idea that all values of σ are equally likely, and we use the density function $f(\sigma) = 1/\sigma$ to formulate the idea that all values of $\log \sigma$ are equally likely. These two prior distributions are not the same. Rather than call (7.2) a "noninformative" distribution, we could think of it as a "standard" or "default" prior distribution.

Strictly speaking, (7.2) is not a probability density function, since its integral, for α and β ranging over all real numbers and σ ranging over all positive numbers, is not 1 but ∞. Therefore (7.2) does not specify a valid probability distribution for the parameters. Nevertheless, if we go ahead and follow the procedure outlined in Section 7.2, we find that the product $f(\alpha, \beta, \sigma) f(\mathbf{y} | \alpha, \beta, \sigma)$ is proportional to a valid probability density function $f(\alpha, \beta, \sigma | \mathbf{y})$, thus yielding a valid posterior distribution.

Bayes Estimates. The Bayes estimates of α and β are taken to be the expectations of α and β under the posterior distribution. These turn out to be exactly the same as the least-squares estimates in (3.2). A verification of this fact is outlined in Section 7.7. Although the Bayes and least-squares estimates are the same, the reasons behind them are very different.

The Rainfall Data. If we admit prior ignorance about the regression line and use (7.2), then the Bayes estimates of α and β are simply the least-squares estimates. Applying formulas (3.2), we estimate the regression line to be $\hat{Y} = 18.03 + 0.5063X$.

Using a Conjugate Prior Distribution. Now suppose we have prior information about the parameters. It is convenient for the purpose of calculation to express this information in terms of a "conjugate" prior distribution. A *conjugate* prior distribution is one that combines with the distribution of the data vector to yield a posterior distribution that has the same form as the prior distribution. When the data vector has a multivariate normal distribution, it turns out that if we let the conditional distribution of (α, β) for each given value of σ be a bivariate normal distribution and let the distribution of $1/\sigma^2$ be a gamma distribution, then this constitutes a conjugate prior distribution for the parameter vector (α, β, σ). By using a conjugate prior distribution we will be able to obtain explicit formulas for the expectations of α and β under the posterior distribution.

Quantifying Prior Information. The decision to use a prior distribution from the family of conjugate distributions is not based on prior information but simply on convenience. Our prior information about the parameters enters the procedure when we select a particular conjugate distribution. For instance, we might quantify our prior information by specifying the expectations, standard deviations and correlation of α and β given σ and the expectation and standard deviation of $1/\sigma^2$. If we are only interested in calculating estimates of α and β, as in this section, it is not necessary to specify the expectation and standard deviation of $1/\sigma^2$.

Rather than translate our prior information into a distribution for α and β, it may be easier, and is equivalent, to translate it into a prior distribution for μ and β, where $\mu = \alpha + \beta x_m$ and x_m is a "middle" x-value. Note that μ is the height of the regression line near the middle of the data. It would often be reasonable to regard our prior information about μ to be independent of our prior information about β, the slope of the regression line. One case in which this assumption of independence is especially suitable is when our prior information comes from a previous least-squares analysis of similar data. In a normal simple linear regression model, if x_m is the average x-value, then $\hat{\mu}_{\text{LS}}$ and $\hat{\beta}_{\text{LS}}$ are independent.

Bayes Estimates. Let $\mu = \alpha + \beta\bar{x}$; that is, take x_m to be the average x-value in the current data set. Suppose that, based on previous information, we specify a conjugate prior distribution in which, conditional on a given value of σ, the expectations of μ and β are e_μ and e_β and their standard deviations are $c_\mu\sigma$ and $c_\beta\sigma$. Let the prior distributions of μ and β be independent. The Bayes estimates of α and β are given by the expectations of α and β under the posterior distribution:

$$\hat{\alpha} = \hat{\mu} - \hat{\beta}\bar{x}$$

$$\hat{\beta} = \left\{ \frac{c_\beta^{-2}}{c_\beta^{-2} + \Sigma(x_i - \bar{x})^2} \right\} e_\beta + \left\{ \frac{\Sigma(x_i - \bar{x})^2}{c_\beta^{-2} + \Sigma(x_i - \bar{x})^2} \right\} \hat{\beta}_{\text{LS}} \qquad (7.3)$$

where

$$\hat{\mu} = \left\{ \frac{c_\mu^{-2}}{c_\mu^{-2} + n} \right\} e_\mu + \left\{ \frac{n}{c_\mu^{-2} + n} \right\} \bar{y}$$

and $\hat{\beta}_{\text{LS}}$ is the least-squares estimate calculated from the current data.

Reasonableness of Formulas (7.3). These formulas show how past knowledge and current data are combined. The Bayes estimate of β is a weighted average of e_β, the prior expected value of β based on past knowledge, and $\hat{\beta}_{LS}$, the least-squares estimate of β based on the current data. The weights are sensible in so far as more weight is put on the current data when there are more data, so that $\Sigma(x_i - \bar{x})^2$ is larger, or when the prior knowledge is more imprecise, that is, c_β is larger. The Bayes estimate of μ is a similar weighted average of the prior expected value based on past knowledge and the least-squares estimate based on the current data. (Note that $\bar{y} = \hat{\mu}_{LS}$.)

The Rainfall Data. Results are available from a previous regression analysis of annual rainfall in Seattle and Portland for the years 1950 through 1979. Let the subscript "0" identify these previous results. The numbers we need for our prior distribution are $\hat{\alpha}_0 = 5.513$, $\hat{\beta}_0 = 0.8961$, $\text{SD}(\hat{\alpha}_0) = 1.140\sigma$, $\text{SD}(\hat{\beta}_0) = 0.02986\sigma$, and the correlation $\text{Corr}(\hat{\alpha}_0, \hat{\beta}_0) = -0.9871$. (These numbers do not completely specify the prior distribution, but they are all we need for the purpose of estimating α and β.) In order to use formulas (7.3), we let $\mu = \alpha + 35.14\beta$, where 35.14 is the average rainfall in Portland for the current data. Then $\hat{\mu}_0 = 5.513 + (35.14)(0.8961) = 37.00$ and $\text{SD}(\hat{\mu}_0) = 0.1977\sigma$. The number 0.1977 is the square root of $(1.140)^2 + (35.14)^2(0.02986)^2 + 2(35.14)(1.140)(0.02986)(-0.9871)$. To specify a prior distribution for the parameters it makes sense to let μ have expectation 37.00 and standard deviation 0.1977σ (conditional on σ) and let β have expectation 0.8961 and standard deviation 0.02986σ. In order to illustrate the use of (7.3) we suppose that the prior distributions of μ and β are independent. Note, however, that the estimates $\hat{\mu}_0$ and $\hat{\beta}_0$ are not exactly independent because the average x-value in the previous data is not exactly 35.14.

Next, from the current data, we calculate $\bar{y} = 35.82$, $\hat{\beta}_{LS} = 0.5063$, and $\Sigma(x_i - \bar{x})^2 = 497.2$. Now we can use formulas (7.3) to obtain the Bayes estimates of α and β. We calculate

$$(0.02986)^{-2}/\left[(0.02986)^{-2} + 497.2\right] = 0.6929$$

$$\hat{\beta} = (0.6929)(0.8961) + (1 - 0.6929)(0.5063) = 0.7764$$

$$(0.1977)^{-2}/\left[(0.1977)^{-2} + 11\right] = 0.6993$$

$$\hat{\mu} = (0.6993)(37.00) + (1 - 0.6993)(35.82) = 36.65$$

$$\hat{\alpha} = 36.65 - (0.7764)(35.14) = 9.363$$

Thus the Bayes estimate of the regression line is $\hat{Y} = 9.363 + 0.7764X$.

7.5 TESTING $\beta = 0$

In the Bayesian approach to testing a hypothesis, rather than calculate a test statistic and its p-value, we calculate the probability, according to the posterior distribution, that the hypothesis is true.

Posterior Probabilities and p-Values. In the classical approach to testing a hypothesis, we calculate a p-value. It is easy to slip into the error of thinking that the p-value is the probability that the null hypothesis is true. Although a p-value can be interpreted as a probability, recall that it is the probability of a different event. Imagine that the experiment is repeated and let D^* denote the event that the data from the repeated experiment yields a test statistic as large or larger than the test statistic obtained from the actual experiment. Let H denote the null hypothesis. The p-value is Prob$(D^*|H)$, the probability that the test statistic in the repeated experiment would be as extreme as the actual observed test statistic, figured under the assumption that H is true. The posterior probability of the null hypothesis, on the other hand, is Prob$(H|D)$, where D denotes the event that the data in the repeated experiment are the same as the actual observed data. So the p-value and the posterior probability both can be regarded as conditional probabilities involving the hypothesis H and the data D but with the conditioning done in opposite directions.

In the classical approach it is not meaningful to ask about the probability of H, and so the reasoning of the test is somewhat indirect. In the Bayesian framework, however, it is possible to directly calculate Prob$(H|D)$. To do this calculation we need to know the distribution of the data vector conditional on the parameter values, which we continue to assume is a multivariate normal distribution, and the prior distribution of the parameters, which we must now specify.

The Prior Distribution. Neither of the two prior distributions that were used in the preceding section to obtain Bayes estimates of the regression line is suitable to use for testing the hypothesis $\beta = 0$. These distributions spread their probability continuously over the parameter set and hence must assign a probability of 0 to any subset of lower dimension. Note that the set $\{(\alpha, \beta, \sigma)$: α any number, $\beta = 0$, $\sigma > 0\}$ of parameters described by the null hypothesis is planar and has dimension 2, whereas the whole parameter set $\{(\alpha, \beta, \sigma)$: α and β any numbers, $\sigma > 0\}$ has dimension 3. Therefore a continuous prior distribution on the parameter set must assign a prior probability of 0 to the null hypothesis, which implies that the posterior probability of the null hypothesis must also be 0. With such a prior distribution we would always conclude $\beta \neq 0$, regardless of the data. So we need a prior distribution that assigns positive probability to the null hypothesis.

In this section we consider only the situation in which no prior knowledge about the parameters is available. We will formulate a prior distribution that seems reasonable in such a situation. Reparameterize the model as

$$y_i = \mu + \beta(x_i - \bar{x}) + e_i \tag{7.4}$$

where $\mu = \alpha + \beta\bar{x}$. The reason for using model (7.4) rather than $y_i = \alpha + \beta x_i + e_i$ is that it is more justifiable to suppose that μ and β are independent than that α and β are (see the subsection on quantifying prior information in Section 7.4). This independence simplifies the presentation of the prior distribution.

The null and alternative hypotheses correspond, respectively, to the sets $H_0 = \{(\mu, \beta, \sigma)$: μ any number, $\beta = 0$, $\sigma > 0\}$ and $H_a = \{(\mu, \beta, \sigma)$: μ any number, $\beta \neq 0$, $\sigma > 0\}$. We can use the following prior distribution:

$$\text{Prob}(H_0) = \tfrac{1}{2}, \qquad \text{Prob}(H_a) = \tfrac{1}{2} \tag{7.5a}$$

$$f(\mu, \sigma | H_0) = 1/\sigma \tag{7.5b}$$

$$f(\mu, \sigma | H_a) = 1/\sigma \tag{7.5c}$$

$$f(\beta | \mu, \sigma, H_a) = ce^{-\beta^2/(2v)} \tag{7.5d}$$

where $c = 1/(\sqrt{2\pi v})$ and $v = n\sigma^2/\Sigma(x_i - \bar{x})^2$.

Justification of the Prior Distribution. In the absence of prior information it seems reasonable to assign equal prior probabilities to the null and alternative hypotheses, as in (7.5a). The density function (7.5b) spreads out the probability $\frac{1}{2}$ over the set H_0 in the same noninformative manner as (7.2).

It makes sense to choose a noninformative prior distribution over the set H_a too. But the fact that H_a involves β in addition to μ and σ complicates the matter. Recall that (7.2) is an improper density function because its integral is not 1 but ∞. So it would be equivalent to let the density function be c/σ for any positive constant c, not necessarily 1. If we chose noninformative distributions similar to (7.2) over both sets H_0 and H_a, their density functions would be $f(\mu, \sigma | H_0) = c_0/\sigma$ and $f(\mu, \beta, \sigma | H_a) = c_a/\sigma$, respectively, but since they are distributions over different sets (even having different dimensions), there is no reason to choose c_0 to be equal to c_a. Unfortunately, the posterior probability of the null hypothesis would very much depend on the relative sizes of c_0 and c_a, and there seems to be no noninformative way to choose them.

So we must come up with a different kind of noninformative distribution for H_a. Ignoring β for the moment, it is reasonable to choose the same noninformative prior distribution for μ and σ when assuming the alternative hypothesis as when assuming the null hypothesis. This is done in (7.5c). It remains to discuss (7.5d).

Justification of (7.5d). The density function in (7.5d) is that of a normal distribution with mean 0 and variance v. The choice of a normal distribution is for convenience. Also, in similar but simpler testing situations it has been found that the shape of the distribution is not crucial. As for the mean of the distribution, since 0 is the hypothesized value of β, it is sensible to have the alternative values distributed around a central value of 0.

Now consider the choice of v. One way we might attempt to obtain a "noninformative" distribution is to let the variance v be very large, so that the probability is spread out over a wide range and the density is almost constant. However, if we let v tend to ∞, the posterior probability of the null hypothesis becomes 1, regardless of the data. Again we run into difficulty in trying to formulate a noninformative prior distribution.

The reason for this difficulty is in the difference between an estimation situation and a testing situation. When we are estimating the value of β, we may sometimes be in a position of complete prior ignorance about its value, but when we are testing the hypothesis $\beta = 0$, then the fact that we have hypothesized the particular value 0 seems to imply at least some prior knowledge or belief. To express a very small amount of prior knowledge, we might suppose that our knowledge is equivalent to the amount of information in a single observation. Being given a single observation still leaves us in ignorance with respect to testing, because it does not provide any basis for deciding between $\beta = 0$ and $\beta \neq 0$. The distance of the observation from 0 could be attributed equally well to either random error or a nonzero value of β, because a single observation gives no information at all about how big the random error might be. Now, what value should be chosen for the variance of the prior distribution of β in order to reflect prior knowledge equivalent to one observation?

We can use the concept of Fisher information. If we assume that the sample $y_1, \ldots, y_n$ comes from a normal distribution and that the value of σ is given, then the amount of Fisher information about β in the sample is equal to the precision (that is, the reciprocal of the variance) of the least-squares estimate of β, which is $\Sigma(x_i - \bar{x})^2/\sigma^2$. This is a reasonable concept of information, since the more precisely we can estimate β, the more information we must have about it. We might say that the amount of Fisher information about β contained in a single observation is one nth part of the information in the whole sample, namely, $\Sigma(x_i - \bar{x})^2/(n\sigma^2)$. Equating information with precision, we take this to be the precision of the prior distribution of β, conditional on σ, when assuming the alternative hypothesis. Its reciprocal is the variance v in (7.5d).

Description of the Test. A Bayesian test of $\beta = 0$ can be performed by calculating the posterior probability of the null hypothesis, based on the prior distribution described in (7.5). The posterior probability can be expressed in terms of the sample correlation between x and y.

$$\text{Prob}(H_0|\mathbf{y}) = \frac{1}{1 + \dfrac{1}{\sqrt{g}}} \tag{7.6}$$

where

$$g = (n+1)\left[1 - \left(\frac{n}{n+1}\right)r^2\right]^{n-1}$$

and r is the correlation coefficient

$$\sum(x_i - \bar{x})(y_i - \bar{y})/\sqrt{\Sigma(x_i - \bar{x})^2\Sigma(y_i - \bar{y})^2}.$$

The posterior probability can also be expressed in terms of the least-squares test statistic t_{LS} in (3.5), because

$$r^2 = \frac{1}{1 + \dfrac{n-2}{t^2}}$$

Justification of Formula (7.6). The formula for the posterior probability can be seen to be reasonable by considering its dependence on the sample correlation coefficient r. We see that $\text{Prob}(H_0|\mathbf{y})$ is a decreasing function of $|r|$. This makes sense, because the larger $|r|$ is, the more highly correlated x and y are, hence the stronger the relationship between them, hence the stronger the evidence against $\beta = 0$, hence the smaller the posterior probability of H_0 should be.

Consider the extreme case when $r = 0$. This indicates that there is no relationship between x and y, which corresponds to $\beta = 0$, and so the posterior probability of H_0 should be close to 1. Indeed, when $r = 0$, $\text{Prob}(H_0|\mathbf{y}) = 1 - 1/(1 + \sqrt{n+1})$. For $n \geq 5$ this is larger than 0.70. At the other extreme, if $r = \pm 1$, this indicates a strong relationship between x and y, which corresponds to $\beta \neq 0$, and so the posterior probability of H_0 should be close to 0. When $r = \pm 1$, $\text{Prob}(H_0|\mathbf{y}) = 1/(1 + (\sqrt{n} - 1)^{n-2})$. For $n \geq 5$ this is smaller than 0.07.

The actual derivation of (7.6) involves some messy integration. A brief outline of the derivation for multiple regression is given in Section 7.8.

The Rainfall Data. Let us test whether there is a significant relationship between the yearly rainfall in Portland and in Seattle. The correlation coefficient is $r = 0.6695$, so $g = 12[1 - \frac{11}{12}(0.6695)^2]^{10} = 0.06044$ and the posterior probability of the null hypothesis is $1/(1 + 1/\sqrt{0.06044}) = 0.1973$. Thus, starting with the noninformative prior distribution (7.5), which assumes that the null hypothesis has a 50% chance of being true, and revising this probability in the light of the data, we conclude that the null hypothesis has only a 20% chance of being true.

7.6 AN EXAMPLE OF MULTIPLE REGRESSION

A study was conducted to assess the effect of several factors on the rate at which a machine can rut asphalt pavement. The factors were: the viscosity of the asphalt, transformed by the logarithm function, (X_1), the percentage of asphalt in the surface (X_2), the percentage of asphalt in the base (X_3), the percentage of fines in the surface (X_4), and the percentage of voids in the surface (X_5). The response variable (Y) was the logarithm of the number of inches of change in the rut depth per million wheel passes. The regression model is $Y = \beta_0 + \beta_1 X_1 + \beta_2 X_2 + \beta_3 X_3 + \beta_4 X_4 + \beta_5 X_5 + e$. Logarithms were taken of the viscosity and the rutting rate based on experience with past data and on plots of the current data. The data are shown in Table 7.2.

Table 7.2 Pavement Data

Identification Number	Log of Rutting Rate (Y)	Log of Viscosity (X_1)	Asphalt (%) in Surface (X_2)	Asphalt (%) in Base (X_3)	Fines (%) (X_4)	Voids (%) (X_5)
1	−0.119	1.944	4.97	4.66	6.5	4.625
2	0.130	1.792	5.01	4.72	8.0	4.977
3	0.158	1.699	4.96	4.90	6.8	4.322
4	0.204	1.763	5.20	4.70	8.2	5.087
5	0.041	1.954	4.80	4.60	6.6	5.971
6	−0.071	1.820	4.98	4.69	6.4	4.647
7	0.079	2.146	5.35	4.76	7.3	5.115
8	−0.252	2.380	5.04	4.80	7.8	5.939
9	−0.143	2.623	4.80	4.80	7.4	5.916
10	−0.328	2.699	4.83	4.60	6.7	5.471
11	−0.481	2.255	4.66	4.72	7.2	4.602
12	−0.585	2.431	4.67	4.50	6.3	5.043
13	−0.119	2.230	4.72	4.70	6.8	5.075
14	−0.097	1.991	5.00	5.07	7.2	4.334
15	0.301	1.544	4.70	4.80	7.7	5.705

Source: Daniel and Wood (1980, p. 110).

7.7 ESTIMATING THE REGRESSION COEFFICIENTS

Let us apply the Bayesian approach to estimating the coefficients in the linear regression model proposed above for the pavement data. Using vector and matrix notation, we let $\boldsymbol{y}$ denote the vector of response variables, $\boldsymbol{X}$ the matrix of explanatory variables, and $\boldsymbol{\beta}$ the vector of regression coefficients. For fixed values of the parameters $\boldsymbol{\beta}$ and σ, assume $\boldsymbol{y}$ has a multivariate normal distribution. As we did in Section 7.4, we try two different prior distributions for the parameters, a noninformative distribution and a conjugate distribution.

Using a Noninformative Prior Distribution. Suppose we have no previous information about pavement rutting rates. As argued in Section 7.4, a prior distribution that can be used to express our ignorance is given by the density function

$$f(\boldsymbol{\beta}, \sigma) = 1/\sigma \tag{7.7}$$

The Bayes Estimate. The Bayes estimate of the parameter vector $\boldsymbol{\beta}$ is the expectation of $\boldsymbol{\beta}$ under the posterior distribution. This expectation turns out to be exactly the same as the least-squares estimate in (3.8).

To see why this is so, we use Bayes's formula. Usually Bayes's formula would be used to obtain the posterior distribution of all the parameters, $\boldsymbol{\beta}$ and σ. But in this section we are interested in $\boldsymbol{\beta}$ and not in σ, and it is simpler to look at the posterior distribution of $\boldsymbol{\beta}$ conditional on σ. We can denote the probability density function of this distribution by $f(\boldsymbol{\beta}|\boldsymbol{y}, \sigma)$. Bayes's formula (7.1), with $\boldsymbol{\beta}$ in place of $\boldsymbol{\theta}$ and with all distributions conditional on σ, says that

$$f(\boldsymbol{\beta}|\boldsymbol{y}, \sigma) = Cf(\boldsymbol{\beta}|\sigma)f(\boldsymbol{y}|\boldsymbol{\beta}, \sigma) \tag{7.8}$$

where C is a quantity that does not involve $\boldsymbol{\beta}$. Since $f(\boldsymbol{\beta}, \sigma)$ does not involve $\boldsymbol{\beta}$, neither does $f(\boldsymbol{\beta}|\sigma)$. Hence $f(\boldsymbol{\beta}|\boldsymbol{y}, \sigma) = Cf(\boldsymbol{y}|\boldsymbol{\beta}, \sigma)$, where C again denotes a (different) quantity that is constant with respect to $\boldsymbol{\beta}$. (Throughout this section, C is used to denote such quantities.) We can determine the distribution of $\boldsymbol{\beta}$ by inspecting $f(\boldsymbol{y}|\boldsymbol{\beta}, \sigma)$.

We are assuming that $f(\boldsymbol{y}|\boldsymbol{\beta}, \sigma)$, regarded as a function of $\boldsymbol{y}$, is the p.d.f. of a multivariate normal distribution with mean vector $\boldsymbol{X\beta}$ and variance–covariance matrix $\sigma^2\boldsymbol{I}$ (where $\boldsymbol{I}$ denotes the identity matrix):

$$f(\boldsymbol{y}|\boldsymbol{\beta}, \sigma) = C\exp\left[\frac{-1}{2\sigma^2}\|\boldsymbol{y} - \boldsymbol{X\beta}\|^2\right] \tag{7.9}$$

(For a vector $\boldsymbol{v}$, the notation $\|\boldsymbol{v}\|$ denotes the length of the vector, that is, $\|\boldsymbol{v}\| = \sqrt{\Sigma v_i^2} = \sqrt{\boldsymbol{v}'\boldsymbol{v}}$.) In a least-squares analysis, the residual vector $\boldsymbol{y} - \boldsymbol{X}\hat{\boldsymbol{\beta}}_{\text{LS}}$ is perpendicular to all the columns of the regression matrix $\boldsymbol{X}$, which implies that $\|\boldsymbol{y} - \boldsymbol{X}\boldsymbol{\beta}\|^2 = \|\boldsymbol{y} - \boldsymbol{X}\hat{\boldsymbol{\beta}}_{\text{LS}}\|^2 + \|\boldsymbol{X}\hat{\boldsymbol{\beta}}_{\text{LS}} - \boldsymbol{X}\boldsymbol{\beta}\|^2$. (This is the Pythagorean theorem.) Therefore

$$f(\boldsymbol{\beta}|\boldsymbol{y},\sigma) = C\exp\left[\frac{-1}{2\sigma^2}\|\boldsymbol{X}\hat{\boldsymbol{\beta}}_{\text{LS}} - \boldsymbol{X}\boldsymbol{\beta}\|^2\right]$$

$$= C\exp\left[\frac{-1}{2\sigma^2}\left(\boldsymbol{\beta} - \hat{\boldsymbol{\beta}}_{\text{LS}}\right)'\boldsymbol{X}'\boldsymbol{X}\left(\boldsymbol{\beta} - \hat{\boldsymbol{\beta}}_{\text{LS}}\right)\right] \tag{7.10}$$

We now see that the p.d.f. of the posterior distribution of $\boldsymbol{\beta}$ conditional on σ is the p.d.f. of a multivariate normal distribution with expectation vector $\hat{\boldsymbol{\beta}}_{\text{LS}}$ and variance–covariance matrix $\sigma^2(\boldsymbol{X}'\boldsymbol{X})^{-1}$.

In particular, the posterior expectation of $\boldsymbol{\beta}$ conditional on σ is the least-squares estimate $\hat{\boldsymbol{\beta}}_{\text{LS}}$. Since it does not depend on σ, it is the posterior expectation of $\boldsymbol{\beta}$ (without conditioning on σ), and hence is the Bayes estimate.

The Pavement Data. If we have no prior knowledge or belief about the regression coefficients in our model, we can use the noninformative prior distribution (7.7). Then the Bayes estimate of $\boldsymbol{\beta}$ is simply the least-squares estimate. Applying formula (3.8) we estimate the regression equation to be $\hat{Y} = -3.362 - 0.5817X_1 + 0.3529X_2 + 0.3831X_3 - 0.009064X_4 + 0.1964X_5$.

Using a Conjugate Prior Distribution. Data are available from a previous study of pavement rutting. This additional information is especially welcome since the size of our sample is somewhat small, only 15. With a small sample, regression estimates are apt to be imprecise. The Bayesian approach allows us to incorporate previous information, which essentially has the effect of increasing the size of the sample and hence increasing the precision of the regression estimates.

The previous data were collected under different conditions and so it may not be appropriate to simply combine both data sets, but the previous study does give us some idea of what might be expected from the current study. We must express the information provided by the previous study in the form of a prior distribution for the parameters.

For the purpose of estimating $\boldsymbol{\beta}$, it is not necessary to specify the entire prior distribution of $\boldsymbol{\beta}$ and σ; it is enough to give the prior distribution of $\boldsymbol{\beta}$ conditional on σ. After we specify a prior p.d.f. $f(\boldsymbol{\beta}|\sigma)$, we will combine it with the p.d.f. (7.9) of the data vector, by means of formula (7.8), to obtain the posterior p.d.f. of $\boldsymbol{\beta}$ conditional on σ. With this in mind, it is convenient to choose the prior p.d.f. to have a form that combines nicely with (7.9).

Let us take the form of the prior distribution of $\boldsymbol{\beta}$ conditional on σ to be multivariate normal; that is,

$$f(\boldsymbol{\beta}|\sigma) = C \exp\left[\frac{-1}{2\sigma^2}(\boldsymbol{\beta} - \boldsymbol{b})'\boldsymbol{V}^{-1}(\boldsymbol{\beta} - \boldsymbol{b})\right] \tag{7.11}$$

with some mean vector $\boldsymbol{b}$ and a variance–covariance matrix $\sigma^2\boldsymbol{V}$ proportional to σ^2. We will choose $\boldsymbol{b}$ and $\boldsymbol{V}$ to reflect our prior knowledge.

Specifying the Prior Information. Based on previous information we want to specify a mean vector and variance–covariance matrix for the prior distribution of $\boldsymbol{\beta}$. For the mean vector $\boldsymbol{b}$ we can choose the estimate of $\boldsymbol{\beta}$ obtained from the previous study, which is

$$\boldsymbol{b} = (-3.55, -0.44, 0.64, 0.13, 0.041, 0.14) \tag{7.12}$$

For the variance–covariance matrix $\sigma^2\boldsymbol{V}$, we can base our choice of $\boldsymbol{V}$ on the estimated variance–covariance matrix of the estimate of $\boldsymbol{\beta}$ from the previous study. Let $\boldsymbol{W}$ denote this latter matrix. We might, for instance, take $\boldsymbol{V} = a\boldsymbol{W}/s^2$, where s^2 is the estimated variance of the random errors in the previous study, and a is a factor greater than 1, which reflects our uncertainty about how directly relevant the previous study is. Suppose we take $a = 4$. Then

$$\boldsymbol{V} = \begin{bmatrix} 1690 & -38 & -174 & -110 & -10 & -47 \\ -38 & 4.07 & 3.46 & 3.23 & 0.47 & 0.23 \\ -174 & 3.46 & 19.81 & 8.88 & 1.15 & 5.31 \\ -110 & 3.23 & 8.88 & 12.47 & -0.17 & 1.56 \\ -10 & 0.47 & 1.15 & -0.17 & 0.60 & 0.15 \\ -47 & 0.23 & 5.31 & 1.56 & 0.15 & 2.69 \end{bmatrix} \tag{7.13}$$

The Bayes Estimate. The Bayes estimate of the parameter vector $\boldsymbol{\beta}$ is the expectation of $\boldsymbol{\beta}$ under the posterior distribution. This expectation is

$$\hat{\boldsymbol{\beta}}_{\text{Bayes}} = \boldsymbol{V}_*\boldsymbol{V}^{-1}\boldsymbol{b} + \boldsymbol{V}_*\boldsymbol{X}'\boldsymbol{X}\hat{\boldsymbol{\beta}}_{\text{LS}} \tag{7.14}$$

where

$$\boldsymbol{V}_* = (\boldsymbol{V}^{-1} + \boldsymbol{X}'\boldsymbol{X})^{-1}$$

Justification of (7.14). The posterior distribution of $\boldsymbol{\beta}$ conditional on σ can be obtained as follows. Since our prior distribution is different, (7.10) is no longer valid for $f(\boldsymbol{\beta}|\mathbf{y}, \sigma)$, but it is still a valid expression for $f(\mathbf{y}|\boldsymbol{\beta}, \sigma)$. We multiply this by (7.11) in accordance with formula (7.8). The product can be manipulated, using the technique of "completing the square", to obtain

$$f(\boldsymbol{\beta}|\mathbf{y}, \sigma) = C \exp\left[\frac{-1}{2\sigma^2}(\boldsymbol{\beta} - \boldsymbol{b}_*)' \boldsymbol{V}_*^{-1}(\boldsymbol{\beta} - \boldsymbol{b}_*)\right]$$

where $\boldsymbol{b}_*$ is the vector displayed in (7.14). This shows that the posterior distribution of $\boldsymbol{\beta}$ conditional on σ is multivariate normal with mean vector $\boldsymbol{b}_*$ and variance–covariance matrix $\sigma^2 \boldsymbol{V}_*$.

Therefore the posterior expectation of $\boldsymbol{\beta}$ conditional on σ is $\boldsymbol{b}_*$. Since it does not depend on σ, it is also the posterior expectation of $\boldsymbol{\beta}$ unconditionally, and hence is the Bayes estimate.

Formula (7.14) can be interpreted as a weighted average of the mean of the prior distribution for $\boldsymbol{\beta}$ and the least-squares estimate of $\boldsymbol{\beta}$, where the "weights" are matrices. The weight matrices are $\boldsymbol{V}_* \boldsymbol{V}^{-1}$ and $\boldsymbol{V}_* \boldsymbol{X}'\boldsymbol{X}$, respectively. Whereas ordinary weights are nonnegative numbers whose sum is 1, these weight matrices are matrices with nonnegative eigenvalues and their sum is the identity matrix. The Bayes estimate is weighted more heavily toward the least-squares estimate if $\boldsymbol{X}'\boldsymbol{X}$ is "large" or if $\boldsymbol{V}^{-1}$ is "small". This makes sense, because $\boldsymbol{X}'\boldsymbol{X}$ is large when the variance–covariance matrix $\sigma^2(\boldsymbol{X}'\boldsymbol{X})^{-1}$ of $\hat{\boldsymbol{\beta}}_{\text{LS}}$ is small, that is, when the least-squares estimate is precise; and $\boldsymbol{V}^{-1}$ is small when $\sigma^2 \boldsymbol{V}$ is large, that is, when our prior information about $\boldsymbol{\beta}$ is imprecise.

One way to express prior ignorance about the regression coefficients would be to let the prior variance–covariance matrix be infinitely large or, in other words, to let $\boldsymbol{V}^{-1}$ be $\mathbf{0}$. Then $\hat{\boldsymbol{\beta}}_{\text{Bayes}} = \hat{\boldsymbol{\beta}}_{\text{LS}}$, which is the same estimate obtained using the noninformative prior distribution (7.7).

The Pavement Data. For the prior distribution we take the distribution specified by (7.11), (7.12), and (7.13). Using formula (7.14) we calculate the Bayes estimate of the regression equation to be $\hat{Y} = -2.713 - 0.5748X_1 + 0.4410X_2 + 0.1706X_3 + 0.002078X_4 + 0.1636X_5$.

7.8 TESTING $\boldsymbol{\beta}_{q+1} = \cdots = \boldsymbol{\beta}_p = 0$

Consider the general linear regression model with p explanatory variables. We will perform a Bayesian test of the hypothesis that $\beta_{q+1} = \cdots = \beta_p = 0$.

The Prior Distribution. Suppose we have no prior knowledge about the parameters. This calls for a noninformative prior distribution. However, the noninformative prior distribution that is used in the preceding section for estimating $\boldsymbol{\beta}$ is not suitable to use for testing it. The reasons for this are discussed in Section 7.5.

An appropriate prior distribution is easier to describe if we first reparameterize so as to achieve orthogonality between the explanatory variables that are included in the hypothesized submodel and those that are excluded. Let $\boldsymbol{W}$ be the $n \times (q+1)$ matrix of explanatory variables $x_{i1}, \ldots, x_{iq}$ included in the hypothesized submodel, together with a first column consisting of 1's, and let $\boldsymbol{Z}$ be the $n \times (p-q)$ matrix of explanatory variables $x_{i,q+1}, \ldots, x_{ip}$ excluded from the hypothesized submodel. Thus $\boldsymbol{X} = (\boldsymbol{W}, \boldsymbol{Z})$. Let $\boldsymbol{\gamma} = (\beta_0, \ldots, \beta_q)$ and $\boldsymbol{\delta} = (\beta_{q+1}, \ldots, \beta_p)$. Then the model $\boldsymbol{y} = \boldsymbol{X\beta} + \boldsymbol{e}$ can be written as

$$\boldsymbol{y} = \boldsymbol{W\gamma} + \boldsymbol{Z\delta} + \boldsymbol{e} \tag{7.15}$$

The hypothesis is that $\boldsymbol{\delta} = \boldsymbol{0}$. The columns of $\boldsymbol{W}$ are not necessarily orthogonal to those of $\boldsymbol{Z}$, but if we define $\boldsymbol{U} = \boldsymbol{Z} - \boldsymbol{W}(\boldsymbol{W}'\boldsymbol{W})^{-1}\boldsymbol{W}'\boldsymbol{Z}$, then $\boldsymbol{W}$ and $\boldsymbol{U}$ are orthogonal. Model (7.15) is equivalent to the model $\boldsymbol{y} = \boldsymbol{W\rho} + \boldsymbol{U\delta} + \boldsymbol{e}$, where $\boldsymbol{\rho} = \boldsymbol{\gamma} + (\boldsymbol{W}'\boldsymbol{W})^{-1}\boldsymbol{W}'\boldsymbol{Z\delta}$.

The following prior distribution is a generalization of distribution (7.5):

$$\text{Prob}(H_0) = \tfrac{1}{2}, \qquad \text{Prob}(H_a) = \tfrac{1}{2} \tag{7.16a}$$

$$f(\boldsymbol{\rho}, \sigma | H_0) = 1/\sigma \tag{7.16b}$$

$$f(\boldsymbol{\rho}, \sigma | H_a) = 1/\sigma \tag{7.16c}$$

$$f(\boldsymbol{\delta} | \boldsymbol{\rho}, \sigma, H_a) = C \exp\left[\frac{-1}{2\sigma^2}\boldsymbol{\delta}'\boldsymbol{V}^{-1}\boldsymbol{\delta}\right] \tag{7.16d}$$

where C is a quantity not involving $\boldsymbol{\delta}$, and $\boldsymbol{V} = n(\boldsymbol{U}'\boldsymbol{U})^{-1}$.

Justification of the Prior Distribution. Distribution (7.16) is intended to reflect prior ignorance about the parameters. In (7.16a) equal prior probabilities are assigned to the null and alternative hypotheses. Within each hypothesis, in (7.16b) and (7.16c), the prior distribution of $\boldsymbol{\rho}$ and σ has the same noninformative form as (7.2). In particular, $\boldsymbol{\rho}$ has a "uniform" distribution (but its density function is improper since its integral is infinite). As discussed in Section 7.5, it is not possible to also give $\boldsymbol{\delta}$ a "uniform" distribution. Instead, we take the viewpoint that some amount of prior knowledge about $\boldsymbol{\delta}$ is implicit in the fact that we have chosen to test the hypothesis $\boldsymbol{\delta} = \boldsymbol{0}$. To express this, we suppose that we have prior knowledge that is equivalent to the amount of information in a single observation.

In (7.16d) the distribution of $\boldsymbol{\delta}$ under the alternative hypothesis, conditional on given values of $\boldsymbol{\rho}$ and σ, is specified to be the multivariate normal distribution with mean vector $\mathbf{0}$ and variance–covariance matrix $\sigma^2 V$. It is sensible to let the mean vector be $\mathbf{0}$ so that the alternative values of $\boldsymbol{\delta}$ are centered around the hypothesized value. The choice of V to be $n(U'U)^{-1}$ can be justified by considering the concept of Fisher information.

If we assume that the data vector $\boldsymbol{y}$ comes from a multivariate normal distribution and that the value of σ is given, then the Fisher information about $\boldsymbol{\delta}$ in the data is equal to the precision matrix (that is, the inverse of the variance–covariance matrix) of the least-squares estimate of $\boldsymbol{\delta}$, which is $U'U/\sigma^2$. Since this represents the information in the whole sample, to represent the information in a single observation we divide by n to get $U'U/(n\sigma^2)$. This can be taken to be the precision matrix for our prior distribution for $\boldsymbol{\delta}$, conditional on σ, when assuming the alternative hypothesis. Its inverse is the variance–covariance matrix $\sigma^2 V$ in (7.16d).

Description of the Test. A Bayesian test of $\beta_{q+1} = \cdots = \beta_p = 0$ is performed by calculating the posterior probability of the null hypothesis. Using the prior distribution (7.16) for the parameters, we obtain the posterior probability

$$\text{Prob}(H_0 | \boldsymbol{y}) = \frac{1}{1 + \dfrac{1}{\sqrt{g}}} \tag{7.17}$$

where

$$g = (n+1)^{p-q}\left[1 - \left(\frac{n}{n+1}\right)R^2\right]^{n-q-1}$$

$$R^2 = \frac{\text{SSR}_{\text{reduced}} - \text{SSR}_{\text{full}}}{\text{SSR}_{\text{reduced}}}$$

SSR_{full} is the sum of squared least-squares residuals for the full model $Y = \beta_0 + \beta_1 X_1 + \cdots + \beta_p X_p + e$ and $\text{SSR}_{\text{reduced}}$ is the sum of squared least-squares residuals for the reduced model $Y = \beta_0 + \beta_1 X_1 + \cdots + \beta_q X_q + e$.

The posterior probability can also be expressed in terms of the least-squares test statistic F_{LS} in (3.13), because

$$R^2 = \frac{1}{1 + \left(\dfrac{n-p-1}{p-q}\right)\dfrac{1}{F_{\text{LS}}}} \tag{7.18}$$

Justification of Formula (7.17). The sum of squared residuals $\text{SSR}_{\text{reduced}}$ measures the amount of variation in the response variable Y that is left unexplained by the explanatory variables $X_1, \ldots, X_q$. So R^2 is the proportion of this unexplained variation that becomes explainable by adding the variables $X_{q+1}, \ldots, X_p$ to the model. Thus R^2 measures the amount of extra information that $X_{q+1}, \ldots, X_p$ provide about the variability of Y beyond the information already provided by $X_1, \ldots, X_q$. Note that $\text{Prob}(H_0|\mathbf{y})$ is a decreasing function of R^2. This makes sense, because the larger R^2 is, the more extra information $X_{q+1}, \ldots, X_p$ provide about the variability of Y, hence the stronger the evidence that $\beta_{q+1}, \ldots, \beta_p$ are not all zero, hence the smaller the posterior probability of H_0 should be.

To formally derive (7.17), we can use Bayes's formula (7.1) with $\boldsymbol{\theta} = (\boldsymbol{\rho}, \boldsymbol{\delta}, \sigma)$, $f(\boldsymbol{\rho}, \boldsymbol{\delta}, \sigma)$ from (7.16), and $f(\mathbf{y}|\boldsymbol{\rho}, \boldsymbol{\delta}, \sigma)$ similar to (7.9) to obtain the posterior p.d.f. $f(\boldsymbol{\rho}, \boldsymbol{\delta}, \sigma|\mathbf{y})$, then integrate the posterior distribution over the set $H_0 = \{(\boldsymbol{\rho}, \boldsymbol{\delta}, \sigma)\colon \boldsymbol{\rho} \text{ any } (q+1)\text{-vector}, \boldsymbol{\delta} = \mathbf{0}, \sigma > 0\}$.

The Pavement Data. Let us test whether the variation in the rutting rates in the pavement data can be explained adequately by the variation in the viscosity of the asphalt. In terms of the parameters, we want to test the hypothesis that $\beta_2 = \beta_3 = \beta_4 = \beta_5 = 0$. We have $n = 15$, $p = 5$, and $q = 1$. The least-squares test statistic is $F_{\text{LS}} = 3.351$. From (7.16) we obtain $R^2 = 1/[1 + (9/4)(1/3.351)] = 0.5983$. Next, $g = (16^4)[1 - (15/16)(0.5983)]^{13} = 1.477$. Now $\text{Prob}(H_0|\mathbf{y}) = 1/(1 + 1/\sqrt{1.477}) = 0.5486$. We have used the noninformative prior distribution (7.16), which assigns probability 0.50 to the null hypothesis. After incorporating the data, we see that the probability of the null hypothesis increases slightly to 0.55. These data are inconclusive about the question of whether the other four explanatory variables have significant explanatory power beyond that of viscosity alone.

7.9 COMPUTATION

One can obtain the least-squares estimate $\hat{\boldsymbol{\beta}}_{\text{LS}}$ and the statistic F_{LS} for testing $\beta_{q+1} = \cdots = \beta_p = 0$ from a wide variety of statistical computer packages. If you want to calculate a Bayesian estimate of $\boldsymbol{\beta}$ using the noninformative prior distribution (7.7), it is simply $\hat{\boldsymbol{\beta}}_{\text{LS}}$. To perform a Bayesian test of $\beta_{q+1} = \cdots = \beta_p = 0$ using the noninformative prior distribution (7.16), the posterior probability of the null hypothesis can be calculated on a hand calculator by putting F_{LS} into formula (7.18) and putting R^2 into (7.17). To incorporate prior information into an estimate of $\boldsymbol{\beta}$, one could use a conjugate prior of the form (7.11). In a matrix-based computer language such as APL or GAUSS or SAS/IML, it would be easy to write a program to calculate the Bayes estimate (7.14). A list of computer programs that are available for doing Bayesian regression is given in Press (1989, Section 3.5).

As a test case for checking the output of a Bayesian regression program, you can use the data in Table 3.2. Doing simple Bayesian regression with $X = X_1$ using the conjugate prior distribution described in Section 7.4 with $e_\mu = 40$, $c_\mu = 0.5$, $e_\beta = 1.2$, and $c_\beta = 0.1$, one obtains $\hat{Y} = 32.65 + 1.293X$ as the estimated regression line. For the noninformative prior distribution (7.5), the posterior probability of the hypothesis $\beta = 0$ is 0.2251. Doing multiple Bayesian regression with both explanatory variables using the conjugate prior distribution described in Section 7.7 with $\boldsymbol{b} = (40, 0, -0.5)$ and

$$V = \begin{bmatrix} 100 & -4 & -2 \\ -4 & 0.3 & 0.1 \\ -2 & 0.1 & 0.1 \end{bmatrix}$$

one obtains $\hat{\boldsymbol{\beta}}_{\text{Bayes}} = (50.91, 0.5687, -0.5846)$. For the noninformative prior distribution (7.16), the posterior probability of the hypothesis $\beta_1 = \beta_2 = 0$ is 0.2899.

NOTES

7.1. Thomas Bayes was a clergyman who lived in England from about 1701 to 1761. He was elected a Fellow of the Royal Society in recognition of his mathematical learning. After his death, a paper containing the result now called Bayes's theorem was found among his belongings and was published in 1764. There is some evidence, however, (see Stigler, 1983) that the theorem may have been discovered earlier by someone else.

7.2a. The prior distribution of $\boldsymbol{\theta}$ can be regarded as a quantification of the data analyst's information and beliefs about the parameters prior to observing the data. The prior distribution could be a summary of previous data or it could be based largely on subjective belief. The usual "frequency" definition of probability, in which the probability of an outcome is taken to be the long-run frequency of the outcome in a long series of repetitions of the same situation, does not apply to a probability distribution that is based on beliefs. Instead, probability is taken to be one's degree of personal belief about whether the outcome will occur. See Barnett (1982, Chapter 3).

Another possible view of the prior distribution is that the parameter vector $\boldsymbol{\theta}$ is truly a random variable and is generated by some random mechanism. In this case one could take the frequency view of probability.

7.2b. Formula (7.1) is an extension of Bayes's original theorem. The basic result is that, for two events A and B, the conditional probability of B given A can be obtained as $\text{Prob}(B|A) = \text{Prob}(B)\text{Prob}(A|B)/\text{Prob}(A)$. This can be derived by noting that $\text{Prob}(A)\text{Prob}(B|A) = \text{Prob}(A \text{ and } B) = \text{Prob}(B \text{ and } A) = \text{Prob}(B)\text{Prob}(A|B)$ and solving for $\text{Prob}(B|A)$. An extension of the basic result says that, for two jointly continuously distributed random vectors $\mathbf{y}$ and $\boldsymbol{\theta}$, the probability density function of the conditional distribution of $\boldsymbol{\theta}$ given $\mathbf{y}$ can be obtained as $f(\boldsymbol{\theta}|\mathbf{y}) = f(\boldsymbol{\theta})f(\mathbf{y}|\boldsymbol{\theta})/f(\mathbf{y})$. We can unambiguously write $f(\boldsymbol{\theta}|\mathbf{y}) = Cf(\boldsymbol{\theta})f(\mathbf{y}|\boldsymbol{\theta})$, saying only that C does not involve $\boldsymbol{\theta}$, because $f(\boldsymbol{\theta}|\mathbf{y})$ must integrate to 1 and so, if needed, C can be obtained as $C = 1/\int f(\boldsymbol{\theta})f(\mathbf{y}|\boldsymbol{\theta})d\boldsymbol{\theta}$.

7.2c. Rather than use the mean of the posterior distribution to estimate $\boldsymbol{\theta}$, it would be equally sensible to use the mode, that is, the value of $\boldsymbol{\theta}$ that maximizes $f(\boldsymbol{\theta}|\mathbf{y})$. Any other measure of the "center" of the posterior distribution would also be sensible. Berger (1985, p. 134) states that, at least for estimating a one-dimensional parameter, the mean is frequently a better estimate than the mode. For a unimodal symmetric distribution like the normal distribution, the mean and mode coincide.

7.4a. For what to do with data whose distribution cannot be assumed to be normal, see Note 7.7a.

7.4b. The prior distribution (7.2) was proposed by Jeffreys (1961, Section 3.1). It is equivalent to saying that the distributions of α, β, and σ are independent with densities $f(\alpha) = 1$, $f(\beta) = 1$, and $f(\sigma) = 1/\sigma$. (The letter f is used generically here to denote any density function. To be more precise we could write the three density functions as $f_\alpha(\alpha)$, $f_\beta(\beta)$, and $f_\sigma(\sigma)$.) The density $f(\alpha) = 1$ represents a "uniform" distribution for α. If the range of α were restricted to a finite interval (a, b) then a uniform distribution for α would be represented by the constant density $f(\alpha) = 1/(b - a)$. There is no proper definition of a uniform distribution on an infinite interval, but nevertheless, for our purposes, letting $f(\alpha)$ be constant over the range of α is sufficient for representing the idea of a uniform distribution for α. So let us choose a positive constant k and define $f(\alpha) = k$ for all α. This is not really a proper probability density function, because the integral of $f(\alpha) = k$ as α ranges over the whole real line is ∞. For $f(\alpha)$ to be proper, the integral would have to be 1. The particular constant k does not matter, as long as it is positive, so we may as well choose the constant 1.

We can show that the constant does not matter with respect to our primary objective, which is to calculate the posterior distribution. Suppose $f(\alpha, \beta, \sigma) = k/\sigma$. To obtain the posterior probability density function we can use Bayes's formula (7.1) with $\boldsymbol{\theta} = (\alpha, \beta, \sigma)$. From Note 7.2b we see that the quantity C in (7.1) is $C = 1/\int\int\int f(\alpha, \beta, \sigma) f(\mathbf{y}|\alpha, \beta, \sigma) d\alpha\, d\beta\, d\sigma$. Thus the constant k appears twice in (7.1) and cancels itself out.

The density $f(\sigma) = 1/\sigma$ is not constant and so does not represent a "uniform" distribution for σ; but consider the parameter $\lambda = \log \sigma$. Its density is $f_\lambda(\lambda) = f_\sigma(\sigma)|d\sigma/d\lambda| = f_\sigma(e^\lambda)e^\lambda = e^{-\lambda}e^\lambda = 1$, and so $\log \sigma$ has a "uniform" distribution.

7.4c. A noninformative prior distribution is called a "reference" prior by Lee (1989, p. 46). It could be thought to represent the view of someone without strong prior beliefs, and results based on it could serve as a reference against which to compare results based on priors expressing diverse personal beliefs. A noninformative prior is also called an "objective" prior by Berger (1985, p. 110). Berger argues that "use of noninformative priors should be recognized as being at least as objective as any other statistical technique."

7.4d. The conjugate prior distribution described in this section is the normal-gamma distribution. See Note 7.7h.

7.4e. The replacement of the parameter α by the parameter $\mu = \alpha + \beta\bar{x}$ is a special case of the replacement of the parameter vector $\boldsymbol{\gamma}$ by the parameter vector $\boldsymbol{\rho} = \boldsymbol{\gamma} + (\boldsymbol{W}'\boldsymbol{W})^{-1}\boldsymbol{W}'\boldsymbol{Z}\boldsymbol{\delta}$ in the subsection on the prior distribution in Section 7.8. In the case of simple regression, $\boldsymbol{\gamma} = \alpha, \boldsymbol{\delta} = \beta, \boldsymbol{W} = \mathbf{1}$, and $\boldsymbol{Z} = \boldsymbol{x}$. Note that $(\mathbf{1}'\mathbf{1})^{-1}\mathbf{1}'\boldsymbol{x} = \bar{x}$.

7.4f. In a least-squares analysis of data following a normal simple linear regression model, the estimates $\hat{\mu}_{\text{LS}}$ and $\hat{\beta}_{\text{LS}}$ are independent, if μ is defined to be $\alpha + \beta\bar{x}$. To see this, first recall from (3.2) that $\hat{\alpha}_{\text{LS}} = \bar{y} - \hat{\beta}_{\text{LS}}\bar{x}$. Therefore $\hat{\mu}_{\text{LS}} = \bar{y}$. In Note 3.4a it is seen that $\hat{\beta}_{\text{LS}} = \Sigma c_i y_i$, where $\Sigma c_i = 0$. Now $\text{Cov}(\bar{y}, \hat{\beta}_{\text{LS}}) = \text{Cov}((1/n)\Sigma_j y_j, \Sigma_i c_i y_i) = (1/n)\Sigma_j \Sigma_i c_i \text{Cov}(y_j, y_i) = (1/n)\Sigma_i c_i \sigma^2 = 0$, because $\text{Cov}(y_j, y_i) = 0$ if $j \neq i$ and $\text{Cov}(y_i, y_i) = \text{Var}(y_i) = \sigma^2$. Since we are assuming that the y_i's are normally distributed, two linear combinations of the y_i's are independent if and only if their covariance is 0.

7.5a. A different Bayesian method of testing $\beta = 0$ can be based on the concept of a highest posterior density (HPD) region. See Box and Tiao (1973, pp. 121–123) and Note 7.8a.

If we choose the noninformative prior distribution (7.2), then the Bayesian test based on an HPD region is exactly the same as the least-squares test based on test statistic t_{LS} in (3.5). That is, the Bayesian test rejects the hypothesis $\beta = 0$ if and only if the p-value of the least-squares test is less than p_0 for some chosen value of p_0.

7.5b. The prior distributions used in Section 7.4 for estimating β are suitable for testing one-sided hypotheses such as $\beta \leq 0$ versus $\beta > 0$. As explained in Section 7.5, these priors are not suitable for testing the precise hypothesis $\beta = 0$ versus $\beta \neq 0$ because the two parameter sets defined by the conditions $\beta = 0$ and $\beta \neq 0$ have different dimensions. Note that the two parameter sets defined by the conditions $\beta \leq 0$ and $\beta > 0$ both have the same dimension 3. See Note 7.5c.

7.5c. A continuous distribution over a set must assign probability 0 to any subset of lower dimension than the set. For example, recall that a continuous distribution on the real line assigns probability 0 to any point. The real line has dimension 1 and a point has dimension 0. Any line segment is one-dimensional, any rectangle (including its interior) is two-dimensional, and any cube (including its interior) is three-dimensional.

7.5d. If the prior probability of the null hypothesis is 0, then so is the posterior probability. To see why this is so, let H denote the null hypothesis and let D denote the data. Bayes's theorem, expressed in terms of probabilities rather than probability density functions, says that $\text{Prob}(H|D) = \text{Prob}(D|H)\text{Prob}(H)/\text{Prob}(D)$. Therefore, if $\text{Prob}(H) = 0$, then it follows that $\text{Prob}(H|D) = 0$.

7.5e. We have chosen the prior probability $\text{Prob}(H_0)$ to be $\frac{1}{2}$, but it is easy to adjust (7.6) to obtain the posterior probability for any choice of $\text{Prob}(H_0)$. First calculate the value of (7.6); call it p_*. This is the posterior probability of H_0 only if $\text{Prob}(H_0) = \frac{1}{2}$. If instead we want to put $\text{Prob}(H_0) = \pi_0$, then the posterior probability of H_0 is $\pi_0 p_* / [\pi_0 p_* + (1 - \pi_0)(1 - p_*)]$.

7.5f. Concerning the arbitrariness of the constant 1 in the density functions (7.5b) and (7.5c), see Note 7.4b.

7.5g. The prior distribution of β, conditional on μ and σ under the alternative hypothesis, is assumed in (7.5d) to be normal with mean 0 and variance v. In the simpler situation of testing the hypothesis $\xi = \xi_0$ for a normal population with unknown mean ξ and known variance, it has been found that the result of the test is not much affected by the shape of the prior distribution for ξ under the alternative hypothesis. See Berger (1985, p. 151) and Lee (1989, p. 138). So presumably the shape of the p.d.f. (7.5d) could be chosen to be Cauchy or uniform, instead of normal, without greatly changing the posterior probability of $\beta = 0$.

The choice of v, however, is crucial. The v in (7.5d) is based on arguments of Jeffreys (1961, Section 5.2). These arguments, when extended to regression (see Zellner and Siow, 1980), say that a good choice for the prior p.d.f. $f(\beta|\mu, \sigma, H_a)$ is Cauchy with scale parameter $\sqrt{v}$.

7.5h. For the definition of Fisher information, see Lehmann (1983, pp. 117–118).

7.5i. Formula (7.6) is a special case of (7.17).

7.7a. The assumption that the distribution of the data vector, given the parameters, is multivariate normal is not always appropriate. In order to obtain closed-form formulas for Bayesian estimates and tests, such as those presented in this chapter, one needs to carefully choose the form of the data distribution and the form of the prior distribution. A much wider choice of distributions is available, however, if we forego closed-form formulas and instead rely on numerical methods.

Suppose we choose probability density functions $f(\boldsymbol{y}|\boldsymbol{\theta})$ and $f(\boldsymbol{\theta})$ for the data and parameters, respectively. If we want an estimate of $\boldsymbol{\theta}$, we can use the mean vector of the posterior distribution of $\boldsymbol{\theta}$, which can be computed as $\int \boldsymbol{\theta} f(\boldsymbol{\theta}) f(\boldsymbol{y}|\boldsymbol{\theta})\, d\boldsymbol{\theta} / \int f(\boldsymbol{\theta}) f(\boldsymbol{y}|\boldsymbol{\theta})\, d\boldsymbol{\theta}$. Numerical methods have been developed for computing such integrals; see Berger (1985, Section 4.9) and Press (1989, Chapter III).

Box and Tiao (1973, Chapter 3) present a class of distributions, which includes the normal, double-exponential, and uniform distributions, for modeling data and show how the posterior distribution can be calculated.

7.7b. Formula (7.8) is a conditional version of Bayes's theorem. It is an extension of the more basic fact that, for three events A, B, and D, it is true that $\text{Prob}(B|A, D) = \text{Prob}(B|D)\,\text{Prob}(A|B, D)/\text{Prob}(A|D)$. This can be derived by noting that $\text{Prob}(D)\,\text{Prob}(A|D)\,\text{Prob}(B|A, D) = \text{Prob}(D \text{ and } A \text{ and } B) = \text{Prob}(D \text{ and } B \text{ and } A) = \text{Prob}(D)\,\text{Prob}(B|D)\,\text{Prob}(A|B, D)$ and solving for $\text{Prob}(B|A, D)$. The extension to probability density functions says that $f(\boldsymbol{\beta}|\boldsymbol{y}, \sigma) = f(\boldsymbol{\beta}|\sigma) f(\boldsymbol{y}|\boldsymbol{\beta}, \sigma)/f(\boldsymbol{y}|\sigma)$. Note that $C = 1/f(\boldsymbol{y}|\sigma)$ does not involve $\boldsymbol{\beta}$.

7.7c. To derive the identity $\|\boldsymbol{y} - \boldsymbol{X\beta}\|^2 = \|\boldsymbol{y} - \boldsymbol{X}\hat{\boldsymbol{\beta}}_{\text{LS}}\|^2 + \|\boldsymbol{X}\hat{\boldsymbol{\beta}}_{\text{LS}} - \boldsymbol{X\beta}\|^2$, let $\boldsymbol{a} = \boldsymbol{y} - \boldsymbol{X\beta}$, $\boldsymbol{b} = \boldsymbol{y} - \boldsymbol{X}\hat{\boldsymbol{\beta}}_{\text{LS}}$, and $\boldsymbol{c} = \boldsymbol{X}\hat{\boldsymbol{\beta}}_{\text{LS}} - \boldsymbol{X\beta}$, and note that $\boldsymbol{a} = \boldsymbol{b} + \boldsymbol{c}$. Also note that $\boldsymbol{b}$ is perpendicular to $\boldsymbol{c}$, because $\boldsymbol{b}$ is the least-squares residual vector and $\boldsymbol{c}$ is a linear combination of the columns of $\boldsymbol{X}$. In the subsection on the geometric interpretation of formula (3.8) in Section 3.7, it is seen that the least-squares residual vector is perpendicular to all the columns of the regression matrix $\boldsymbol{X}$. The Pythagorean theorem says that if $\boldsymbol{b}$ and $\boldsymbol{c}$ are perpendicular, then $\|\boldsymbol{b} + \boldsymbol{c}\|^2 = \|\boldsymbol{b}\|^2 + \|\boldsymbol{c}\|^2$.

7.7d. The p.d.f. (7.10) has the same form as (7.11) with $\boldsymbol{b} = \hat{\boldsymbol{\beta}}_{\text{LS}}$ and $\boldsymbol{V} = (\boldsymbol{X}'\boldsymbol{X})^{-1}$. Therefore it is the p.d.f. of a multivariate normal distribution.

7.7e. In the subsection on the Bayes estimate using a noninformative prior distribution, we show that $E(\boldsymbol{\beta}|\sigma, \boldsymbol{y}) = \hat{\boldsymbol{\beta}}_{\text{LS}}$ for all σ. Let us show that $E(\boldsymbol{\beta}|\boldsymbol{y}) = \hat{\boldsymbol{\beta}}_{\text{LS}}$. For two jointly distributed random variables T and S, recall that the expectation of T can be computed in two stages using conditional expectation. First calculate the conditional expectation of T given $S = s$, that is, $E(T|s)$. Then $E(T|S)$, being a function of the random variable S, is itself a random variable, and its expectation is $E[E(T|S)] = E(T)$. For three jointly distributed random variables T, S, and Y, a similar equation holds:

$E[E(T|S, y)|y] = E(T|y)$. The result extends to jointly distributed random vectors. Therefore $E(\boldsymbol{\beta}|\mathbf{y}) = E[E(\boldsymbol{\beta}|\sigma, \mathbf{y})|\mathbf{y}] = E(\hat{\boldsymbol{\beta}}_{\text{LS}}|\mathbf{y}) = \hat{\boldsymbol{\beta}}_{\text{LS}}$.

7.7f. An important part of Bayesian data analysis is the choice of the prior distribution. After performing a Bayesian analysis, one might wonder how crucially the results depend on the particular prior distribution that was chosen. A way to deal with this question is to perform analyses using several different priors and see how the results differ. See Berger (1985, Section 4.7).

7.7g. It can be difficult to express one's prior knowledge and beliefs in the form of a prior distribution on the parameters. One could have a lot of past experience with data sets similar to the current data set and yet have trouble expressing this experience directly in terms of the parameters in a regression model. It might be more natural to express one's knowledge about the relationship between the response variable Y and the explanatory variables X_j in the form of predictions for the value of Y for various values of the X_j. Procedures have been developed for translating such predictions into the form of a prior distribution for the parameters. See Kadane et al. (1980).

7.7h. A *conjugate prior distribution* for $\boldsymbol{\beta}$ and σ is one such that, after it is combined with the distribution of the data vector to obtain the posterior distribution, the posterior distribution turns out to have the same form as the prior distribution. When the data vector has the multivariate normal distribution (7.9), a conjugate prior distribution is obtained by using the normal-gamma distribution. We let the prior distribution of $\boldsymbol{\beta}$ conditional on σ be multivariate normal as in (7.11) and let the prior distribution of $1/\sigma^2$ be gamma. We can show that the posterior distribution of $\boldsymbol{\beta}$ and σ has the same form.

Define $\psi = 1/\sigma^2$. The prior p.d.f. of ψ is taken to be $f(\psi) = C\psi^{a-1}e^{-\psi/d}$ for positive constants a and d. This is the gamma p.d.f. with parameters a and d. To avoid confusion, we call a and d "hyperparameters". In this note we use the symbol C generically to denote any quantity that involves neither $\boldsymbol{\beta}$ nor ψ. In (7.11) we have regarded σ as a fixed constant, but since we are now concerned with σ as well as $\boldsymbol{\beta}$, we should note that the quantity C in (7.11) involves a factor $\sigma^{-(p+1)}$. The prior p.d.f. of $\boldsymbol{\beta}$ conditional on ψ is $f(\boldsymbol{\beta}|\psi) = C\psi^{(p+1)/2}\exp[(-\psi/2)(\boldsymbol{\beta} - \boldsymbol{b})'V^{-1}(\boldsymbol{\beta} - \boldsymbol{b})]$ for a vector $\boldsymbol{b}$ and a positive definite matrix V. The joint p.d.f. of the parameters is the normal-gamma p.d.f. $f(\boldsymbol{\beta}, \psi) = f(\boldsymbol{\beta}|\psi)f(\psi)$ with hyperparameters $\boldsymbol{b}$, V, a, and d.

In (7.9) we have also regarded σ as a fixed constant, and so now we must rewrite the p.d.f. of the data, conditional on the parameters, as $f(\mathbf{y}|\boldsymbol{\beta}, \psi) = C\psi^{n/2}\exp[(-\psi/2)(\mathbf{y} - X\boldsymbol{\beta})'(\mathbf{y} - X\boldsymbol{\beta})]$. According to Bayes's formula (7.1), the posterior p.d.f. of the parameters is $f(\boldsymbol{\beta}, \psi|\mathbf{y}) = Cf(\boldsymbol{\beta}, \psi)f(\mathbf{y}|\boldsymbol{\beta}, \psi)$. By algebraic manipulation, we can show that this is also a normal-gamma p.d.f. but with different hyperparameters, $\boldsymbol{b}_*$, V_*, a_*, and d_*, where

$b_* = V_*(V^{-1}b + X'y)$, V_* is as in (7.14), $a_* = a + (n/2)$, and $d_* = 1/((1/d) + \frac{1}{2}(y'y + b'V^{-1}b - b_*'V_*'^{-1}b_*))$. Part of the algebraic manipulation is demonstrated in Note 7.7j.

7.7i. The construction of V using an uncertainty factor a is somewhat *ad hoc*. One could try several choices of a and hope that the estimates are not very different.

7.7j. The technique of "completing the square" is used to reexpress $f(\beta|y, \sigma) = C \exp[(-1/2\sigma^2)(\beta - b)'V^{-1}(\beta - b)]\exp[(-1/2\sigma^2)(\beta - \hat{\beta}_{LS})'X'X(\beta - \hat{\beta}_{LS})]$. (In this note the letter C is used to denote any quantity that does not involve β.) Let $A = (\beta - b)'V^{-1}(\beta - b) = \beta'V^{-1}\beta - 2b'V^{-1}\beta + b'V^{-1}b$ and $B = (\beta - \hat{\beta}_{LS})'X'X(\beta - \hat{\beta}_{LS}) = \beta'X'X\beta - 2\hat{\beta}'_{LS}X'X\beta + \hat{\beta}'_{LS}X'X\hat{\beta}_{LS}$. Now

$$
\begin{aligned}
&C \exp\left[\frac{-1}{2\sigma^2}A\right]\exp\left[\frac{-1}{2\sigma^2}B\right] \\
&= C \exp\left[\frac{-1}{2\sigma^2}(A + B)\right] \\
&= C \exp\left[\frac{-1}{2\sigma^2}\{\beta'V^{-1}\beta - 2b'V^{-1}\beta + \beta'X'X\beta - 2\hat{\beta}'_{LS}X'X\beta\}\right] \\
&= C \exp\left[\frac{-1}{2\sigma^2}\left\{\beta'(V^{-1} + X'X)\beta - 2\left(b'V^{-1} + \hat{\beta}'_{LS}X'X\right)\beta\right\}\right] \\
&= C \exp\left[\frac{-1}{2\sigma^2}\{\beta'W\beta - 2h'\beta\}\right]
\end{aligned}
$$

where $W = V^{-1} + X'X$ and $h = V^{-1}b + X'X\hat{\beta}_{LS}$. (In the preceding string of equalities, the C in the second expression is different from the C in the third expression, but neither of them involves β.) Now we "complete the square" by writing $\beta'W\beta - 2h'\beta = (\beta - W^{-1}h)'W(\beta - W^{-1}h) - h'W^{-1}h$. (This generalizes the equality $b^2 - 2hb = (b - h)^2 - h^2$.) Note that $W = V_*^{-1}$ and $W^{-1}h = b_*$, where V_* and $b_* = \hat{\beta}_{\text{Bayes}}$ are as in (7.14). Absorbing the term $\exp[(1/2\sigma^2)h'W^{-1}h]$ into C, we obtain $f(\beta|y, \sigma) = C \exp[(-1/2\sigma^2)(\beta - b_*)'V_*^{-1}(\beta - b_*)]$.

7.7k. The weight matrices in (7.14) are V_*V^{-1} and $V_*X'X$. From the definition of V_* we see that the sum of these two matrices is $V_*(V^{-1} + X'X) = I$, the identity matrix.

The usual definition of a weighted average of two vectors b and c is a vector $a = wb + (1 - w)c$, where $0 \le w \le 1$. Let us call this a scalar-weighted average. This can be generalized to the notion of a *matrix-weighted average*, which is defined to be a vector $a = Wb + (I - W)c$, where W is a matrix having all its eigenvalues between 0 and 1. An *eigenvalue* of $\boldsymbol{W}$ is a number λ such that $\boldsymbol{Wx} = \lambda\boldsymbol{x}$ for some nonzero vector $\boldsymbol{x}$. Note that a scalar-weighted average occurs as the special case in which $\boldsymbol{W} = w\boldsymbol{I}$ and $0 \le w \le 1$.

When $\boldsymbol{W}$ is a diagonal matrix, each component of the matrix-weighted average $\boldsymbol{a}$ is a weighted average of the corresponding components of $\boldsymbol{b}$ and $\boldsymbol{c}$. Specifically, $a_i = w_i b_i + (1 - w_i)c_i$, where w_i is the ith diagonal entry of $\boldsymbol{W}$. The eigenvalues of a diagonal matrix are its diagonal entries.

In general, a matrix-weighted average of two vectors can be expressed as a sum of scalar-weighted averages: $\boldsymbol{a} = \Sigma\boldsymbol{a}_i$, where $\boldsymbol{a}_i = w_i\boldsymbol{b}_i + (1 - w_i)\boldsymbol{c}_i$, $\boldsymbol{b} = \Sigma\boldsymbol{b}_i$, $\boldsymbol{c} = \Sigma\boldsymbol{c}_i$, and the w_i's are the eigenvalues of $\boldsymbol{W}$. A scalar-weighted average is an "average" of $\boldsymbol{b}$ and $\boldsymbol{c}$ in the sense that it lies "between" them; it lies on the line segment joining $\boldsymbol{b}$ and $\boldsymbol{c}$. A matrix-weighted average can be said to lie "between" $\boldsymbol{b}$ and $\boldsymbol{c}$ to the extent that it is contained in a multidimensional parallelogram with $\boldsymbol{b}$ and $\boldsymbol{c}$ at opposite corners.

7.8a. A different Bayesian method of testing $\beta_{q+1} = \cdots = \beta_p = 0$ can be based on the concept of a highest posterior density (HPD) region. See Box and Tiao (1973, Sections 2.8 and 2.9) and Broemeling (1985, pp. 11–14). Let $\boldsymbol{\delta} = (\beta_{q+1}, \ldots, \beta_p)$. Suppose we want to test $\boldsymbol{\delta} = \boldsymbol{0}$ at the 5% level of significance. After obtaining the posterior p.d.f. $f(\boldsymbol{\beta}|\boldsymbol{y})$, one then finds the posterior p.d.f. for $\boldsymbol{\delta}$, $f(\boldsymbol{\delta}|\boldsymbol{y}) = \int f(\boldsymbol{\gamma}, \boldsymbol{\delta}|\boldsymbol{y})\, d\boldsymbol{\gamma}$, where $\boldsymbol{\gamma} = (\beta_0, \ldots, \beta_q)$. Note that $f(\boldsymbol{\gamma}, \boldsymbol{\delta}|\boldsymbol{y}) = f(\boldsymbol{\beta}|\boldsymbol{y})$. An *HPD region* for $\boldsymbol{\delta}$ is a set formed by including, for some fixed positive number c, all parameter vectors $\boldsymbol{\delta}$ such that $f(\boldsymbol{\delta}|\boldsymbol{y}) \ge c$. Let R be an HPD region for $\boldsymbol{\delta}$. We say it is a 95% HPD region if $\text{Prob}(\boldsymbol{\delta} \in R|\boldsymbol{y}) = 0.95$. A test of $\boldsymbol{\delta} = \boldsymbol{0}$ can be performed by rejecting the hypothesis if R does not include $\boldsymbol{0}$.

If we choose the noninformative prior distribution (7.7), then the Bayesian test based on an HPD region is exactly the same as the least-squares test based on test statistic F_{LS} in (3.13). That is, the Bayesian test rejects the hypothesis $\beta_{q+1} = \cdots = \beta_p = 0$ if and only if the p-value of the least-squares test is less than 0.05.

7.8b. To incorporate previous knowledge into the test through an informative prior distribution, see Zellner (1971, Section 10.4).

7.8c. Instead of working with the parameter vectors $\boldsymbol{\gamma}$ and $\boldsymbol{\delta}$, it is convenient to reparameterize to achieve orthogonality. Since we are testing the value of $\boldsymbol{\delta}$, we leave it as it is and we consider transformations of $\boldsymbol{\gamma}$ to $\boldsymbol{\rho} = \boldsymbol{\gamma} + \boldsymbol{A\delta}$ for some matrix $\boldsymbol{A}$. The model $\boldsymbol{y} = \boldsymbol{W\gamma} + \boldsymbol{Z\delta} + \boldsymbol{e}$ becomes $\boldsymbol{y} = \boldsymbol{W}(\boldsymbol{\rho} - \boldsymbol{A\delta}) + \boldsymbol{Z\delta} + \boldsymbol{e} = \boldsymbol{W\rho} + (\boldsymbol{Z} - \boldsymbol{WA})\boldsymbol{\delta} + \boldsymbol{e}$. We want to choose $\boldsymbol{A}$ so that the columns of $\boldsymbol{W}$ are orthogonal to the columns of $\boldsymbol{Z} - \boldsymbol{WA}$, that is, $\boldsymbol{W}'(\boldsymbol{Z} - \boldsymbol{WA}) = \boldsymbol{0}$. Solving for $\boldsymbol{A}$, we obtain $\boldsymbol{A} = (\boldsymbol{W}'\boldsymbol{W})^{-1}\boldsymbol{W}'\boldsymbol{Z}$.

7.8d. For the definitions of Fisher information and information matrix, see Lehmann (183, pp. 117–118, 125–126). If a data vector $\boldsymbol{y}$ has a distribution that depends on parameter vector $\boldsymbol{\theta}$, the information matrix for $\boldsymbol{\theta}$ is approximately equal to the precision matrix of the maximum likelihood estimate of $\boldsymbol{\theta}$ (see Lehmann, 1983, Theorem 6.4.1(ii) and Corollary 2.6.1). We are concerned with the situation in which $\boldsymbol{y}$ has a multivariate normal distribution with mean vector $\boldsymbol{X\beta}$ and variance–covariance matrix $\sigma^2\boldsymbol{I}$. Suppose σ is known. The maximum likelihood estimate of $\boldsymbol{\beta}$ is the same as the least-squares estimate $\hat{\boldsymbol{\beta}}_{\text{LS}}$, and the information matrix for $\boldsymbol{\beta}$ is exactly equal to the precision matrix of the maximum likelihood estimate of $\boldsymbol{\beta}$, namely, $\text{Cov}(\hat{\boldsymbol{\beta}}_{\text{LS}})^{-1} = \boldsymbol{X}'\boldsymbol{X}/\sigma^2$. If we are interested only in the $\boldsymbol{\delta}$ part of $\boldsymbol{\beta}$, it is natural to take the information matrix for $\boldsymbol{\delta}$ to be $\text{Cov}(\hat{\boldsymbol{\delta}}_{\text{LS}})^{-1} = \boldsymbol{U}'\boldsymbol{U}/\sigma^2$.

7.8e. We have chosen the prior probability $\text{Prob}(H_0)$ to be $\frac{1}{2}$, but it is easy to adjust (7.17) to obtain the posterior probability for any choice of $\text{Prob}(H_0)$. Use the formula in Note 7.5e.

7.8f. The quantity R^2 is sometimes called the *coefficient of partial determination*. When $q = 0$, then $\text{SSR}_{\text{reduced}} = \Sigma(y_i - \bar{y})^2$, the total sum of squares (corrected for the mean), and so R^2 is the coefficient of multiple determination. Another description of the coefficient of partial determination is as follows. First, using $X_1, \ldots, X_q$ as explanatory variables, take each of the variables $Y, X_{q+1}, \ldots, X_p$ as the response variable in a linear regression equation, and calculate least-squares estimates. Denote the residuals by $W, Z_{q+1}, \ldots, Z_p$. Then R^2 is the maximum squared correlation between W and a linear combination of $Z_{q+1}, \ldots, Z_p$. See Neter, Wasserman, and Kutner (1989, Section 8.3).

7.8g. Formula (7.17) can be obtained through a derivation similar to that in Zellner and Siow (1980). They used a multivariate Cauchy, rather than normal, prior distribution. To sketch the derivation of (7.17), we begin by expressing the posterior probability of H_0 as

$$\text{Prob}(H_0|\boldsymbol{y}) = \int_{H_0} f(\boldsymbol{\theta}|\boldsymbol{y})\, d\boldsymbol{\theta}$$

where $\boldsymbol{\theta}$ denotes the parameter vector $(\boldsymbol{\rho}, \boldsymbol{\delta}, \sigma)$. By Bayes's formula (see Note 7.2b), $f(\boldsymbol{\theta}|\boldsymbol{y}) = f(\boldsymbol{y}|\boldsymbol{\theta})f(\boldsymbol{\theta})/f(\boldsymbol{y})$, and so $\text{Prob}(H_0|\boldsymbol{y}) = J_0/f(\boldsymbol{y})$ where $J_0 = \int_{H_0} f(\boldsymbol{y}|\boldsymbol{\theta})f(\boldsymbol{\theta})\, d\boldsymbol{\theta}$. Similarly, $\text{Prob}(H_a|\boldsymbol{y}) = J_a/f(\boldsymbol{y})$ where $J_a = \int_{H_a} f(\boldsymbol{y}|\boldsymbol{\theta})f(\boldsymbol{\theta})\, d\boldsymbol{\theta}$. (Just as H_0 denotes the set of parameter vectors under the null hypothesis, H_a denotes the set $\{\boldsymbol{\theta}: \boldsymbol{\rho} \text{ any } (q+1)\text{-vector}, \boldsymbol{\delta} \neq \boldsymbol{0}, \sigma > 0\}$ of parameter vectors under the alternative hypothesis.) Since $\text{Prob}(H_0|\boldsymbol{y}) + \text{Prob}(H_a|\boldsymbol{y}) = 1$, then $J_0 + J_a = f(\boldsymbol{y})$, and so $\text{Prob}(H_0|\boldsymbol{y}) = J_0/(J_0 + J_a)$. It remains to evaluate the integrals defining J_0 and J_a.

We are assuming a normal distribution for the data vector:

$$f(\mathbf{y}|\boldsymbol{\theta}) = (2\pi)^{-n/2}\sigma^{-n}\exp\left[\frac{-1}{2\sigma^2}(\mathbf{y} - \mathbf{W}\boldsymbol{\rho} - \mathbf{U}\boldsymbol{\delta})'(\mathbf{y} - \mathbf{W}\boldsymbol{\rho} - \mathbf{U}\boldsymbol{\delta})\right]$$

Under the null hypothesis, $f(\boldsymbol{\theta}) = f(\boldsymbol{\rho}, \mathbf{0}, \sigma) = \text{Prob}(H_0)f(\boldsymbol{\rho}, \sigma|H_0) = \frac{1}{2}\sigma^{-1}$. Hence $J_0 = \int\int \frac{1}{2}(2\pi)^{-n/2}\sigma^{-n-1}\exp[(-1/2\sigma^2)(\mathbf{y} - \mathbf{W}\boldsymbol{\rho})'(\mathbf{y} - \mathbf{W}\boldsymbol{\rho})]\,d\boldsymbol{\rho}\,d\sigma$. Although the details are messy, the basic idea that allows us to evaluate this integral is simple. If an integrand can be rearranged into the form of a p.d.f. multiplied by a constant factor, then, since a p.d.f. must integrate to 1, the integral is equal to the constant factor. In particular, letting σ be fixed, we find that the integrand of J_0 as a function of $\boldsymbol{\rho}$ is the p.d.f. of a multivariate normal distribution multiplied by a factor that is constant as far as $\boldsymbol{\rho}$ is concerned. This takes care of integration with respect to $\boldsymbol{\rho}$. Now J_0 is in the form of an integral with respect to σ. By changing the variable of integration from σ to $t = \frac{1}{2}\sigma^{-2}$, the integrand can be written as the p.d.f. of a gamma distribution multiplied by a constant factor. The same approach works for evaluating J_a.

Additional Reading. For an introduction to Bayesian statistics, see Zellner (1971, Chapter II), Box and Tiao (1973, Chapter 1), Berger (1985, Chapter 4), Lee (1989), or Press (1989, Chapters 1 and 2). For more on Bayesian regression, see Zellner (1971, Chapter III), Box and Tiao (1973, Section 2.7), Broemeling (1985, Chapter 1 and pp. 84–104), Lee (1989, Sections 6.3, 6.4, and 6.7), and Press (1989, Chapter V).

REFERENCES

Barnett, V. (1982). *Comparative Statistical Inference*, 2nd ed. Wiley, New York.

Berger, J. O. (1985). *Statistical Decision Theory and Bayesian Analysis*, 2nd ed. Springer-Verlag, New York.

Box, G. E. P., and G. C. Tiao (1973). *Bayesian Inference in Statistical Analysis*. Addison-Wesley, Reading, MA.

Broemeling, L. D. (1985). *Bayesian Analysis of Linear Models*. Marcel Dekker, New York.

Daniel, C., and F. S. Wood (1980). *Fitting Equations to Data*, 2nd ed. Wiley, New York.

Jeffreys, H. (1961). *Theory of Probability*, 3rd ed. (1st ed. 1939). Oxford University Press, London.

Kadane, J. B., J. M. Dickey, R. L. Winkler, W. S. Smith, and S. C. Peters (1980). Interactive elicitation of opinion for a normal linear model. *Journal of the American Statistical Association*, vol. 75, pp. 845–854.

Lee, P. M. (1989). *Bayesian Statistics: An Introduction*. Oxford University Press, New York.

Lehmann, E. L. (1983). *Theory of Point Estimation*. Wiley, New York.

Neter, J., W. Wasserman, and M. H. Kutner (1989). *Applied Linear Regression Models*, 2nd ed. Irwin, Homewood, IL.

National Oceanic and Atmospheric Administration (1990). Local Climatological Data. NOAA, Asheville, NC.

Press, S. J. (1989). *Bayesian Statistics: Principles, Models and Applications*. Wiley, New York.

Stigler, S. M. (1986). *The History of Statistics: The Measurement of Uncertainty Before 1900*. Harvard University Press, Cambridge, MA.

Stigler, S. M. (1983). Who discovered Bayes's theorem? *American Statistician*, vol. 37, pp. 290–296.

Zellner, A. (1971). *An Introduction to Bayesian Inference in Econometrics*. Wiley, New York.

Zellner, A., and A. Siow (1980). Posterior odds ratios for selected regression hypotheses. In J. M. Bernardo, M. H. DeGroot, D. V. Lindley, and A. F. M. Smith (eds.), *Bayesian Statistics*. University Press, Valencia.

CHAPTER 8

Ridge Regression

8.1 INTRODUCTION

Ridge regression was introduced in 1962 by Arthur Hoerl in an article in a chemical engineering journal. In his experience in regression analysis he had found that when there were correlations among the explanatory variables, the least-squares estimates often did not make sense when put into the context of the process that generated the data. He proposed a method to obtain better estimates.

In simple regression with only one explanatory variable, of course, there can be no problem with correlations among the explanatory variables, but it is nevertheless instructive to look at the case of simple regression to explain how ridge regression attempts to produce more accurate estimates.

Ridge regression is concerned only with estimation, and so testing is not considered in this chapter.

8.2 AN EXAMPLE OF SIMPLE REGRESSION

In Table 8.1 are the heights, in centimeters, of 32 girls at ages 9 and 18. The plot in Figure 8.1 shows that there is an approximately linear relationship between the heights at the two ages. Let us use ridge regression to estimate height at age 18 as a linear function of height at age 9.

8.3 ESTIMATING THE REGRESSION LINE

Standardization. The first step in ridge regression is usually to standardize the explanatory variables. The reasons for standardization are given in Section 8.5. To standardize the variable X, let $z_i = (x_i - \bar{x})/s_x$, where $\bar{x}$ is the average of the x_i's and s_x is the standard deviation of the x_i's. For the

Table 8.1 Height Data

Identification Number of Girl	Height at Age 18 (Y)	Height at Age 9 (X)	Standardized Height at Age 9 (Z)
1	169.6	136.5	0.2154
2	166.8	137.0	0.3033
3	157.1	129.0	−1.1034
4	181.1	139.4	0.7254
5	158.4	125.6	−1.7013
6	165.6	137.1	0.3209
7	166.7	133.6	−0.2945
8	156.5	121.4	−2.4398
9	168.1	133.6	−0.2945
10	165.3	134.1	−0.2066
11	163.7	139.4	0.7254
12	173.7	138.1	0.4968
13	169.2	138.4	0.5495
14	170.1	139.5	0.7429
15	164.2	129.8	−0.9627
16	176.0	144.8	1.6749
17	170.9	138.9	0.6374
18	169.2	140.3	0.8836
19	172.0	143.8	1.4991
20	163.0	133.6	−0.2945
21	154.5	123.5	−2.0706
22	172.5	139.9	0.8133
23	175.6	136.1	0.1451
24	167.2	135.8	0.0923
25	164.0	131.9	−0.5935
26	161.6	130.9	−0.7693
27	153.6	126.3	−1.5782
28	173.5	135.5	0.0396
29	166.2	134.0	−0.2242
30	162.8	138.2	0.5143
31	168.6	142.0	1.1825
32	169.2	140.8	0.9715

Source: Weisberg (1985, p. 57).

height data, $\bar{x} = 135.275$ and $s_x = 5.687$, keeping extra significant digits in $\bar{x}$ to avoid too much round-off error. Hence, for example, $z_1 = (136.5 - 135.275)/5.687 = 0.2154$.

In terms of the unstandardized variable X, the simple linear regression model is $Y = \alpha + \beta X + e$. After standardizing, let us write $Y = \mu + \gamma Z + e$.

The Ridge Estimates. The least-squares estimates of μ and γ can be obtained from formulas (3.2). Since the z_i's are standardized, we have $\bar{z} = 0$

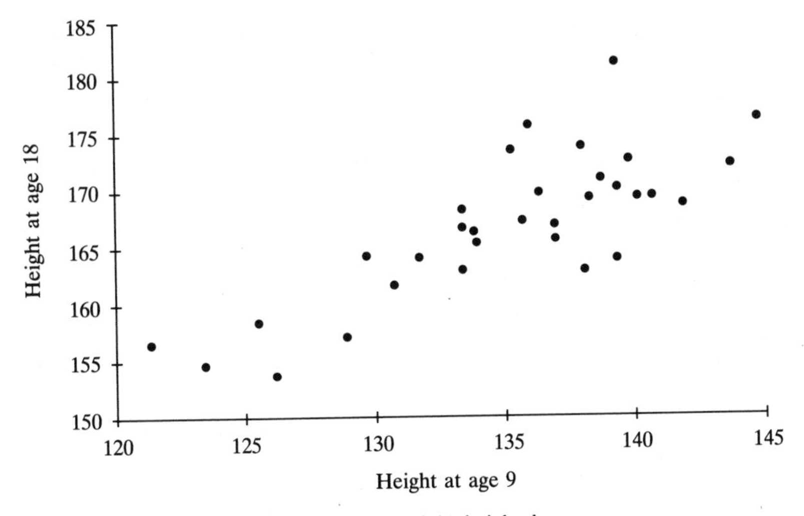

Figure 8.1 Plot of the height data.

and so the formulas simplify to

$$\hat{\gamma}_{\text{LS}} = \frac{\Sigma z_i y_i}{\Sigma z_i^2}$$

$$\hat{\mu}_{\text{LS}} = \bar{y}$$

For the ridge estimates, we keep $\hat{\mu} = \bar{y}$ and let

$$\hat{\gamma} = \frac{\Sigma z_i y_i}{\Sigma z_i^2 + k} \tag{8.1}$$

where $k = \hat{\sigma}_{\text{LS}}^2 / \hat{\gamma}_{\text{LS}}^2$ and $\hat{\sigma}_{\text{LS}}$ is the least-squares estimate of σ in formula (3.3).

Mean Squared Error. Before explaining how formula (8.1) might produce a more accurate estimate than $\hat{\gamma}_{\text{LS}}$, we need to specify what we mean by "accurate". Roughly speaking, $\hat{\gamma}$ is accurate if, viewed as a random variable, it tends to be close to γ, that is, if the distance between $\hat{\gamma}$ and γ tends to be small. The most common way to quantify this idea is to measure the accuracy of an estimate $\hat{\gamma}$ by its *mean squared error*, $\text{MSE}(\hat{\gamma}) = E[(\hat{\gamma} - \gamma)^2]$, the expectation of the squared distance between $\hat{\gamma}$ and γ. The smaller $\text{MSE}(\hat{\gamma})$ is, the more accurate $\hat{\gamma}$ is.

A useful expression is

$$\text{MSE}(\hat{\gamma}) = \text{Var}(\hat{\gamma}) + [E(\hat{\gamma}) - \gamma]^2 \tag{8.2}$$

That is, the mean squared error of $\hat{\gamma}$ is equal to its variance plus the square of its bias. So, for an estimate to have small mean squared error, it must simultaneously have small variance and small bias.

The least-squares estimate $\hat{\gamma}_{LS}$ is unbiased; that is, $E(\hat{\gamma}_{LS}) = \gamma$. The ridge estimate is biased but it is intended to have smaller variance. The hope is that the increase in squared bias is less than the decrease in variance, so that the ridge estimate will have smaller mean squared error.

Justification of (8.1). One way to think about ridge estimation is that it tries to improve the accuracy of $\hat{\gamma}_{LS}$ by "shrinking" it. To shrink $\hat{\gamma}_{LS}$, multiply it by a number c between 0 and 1 to obtain $\hat{\gamma} = c\hat{\gamma}_{LS}$. We call this "shrinking" because $|c\hat{\gamma}_{LS}| < |\hat{\gamma}_{LS}|$.

The mean squared error of $\hat{\gamma}_{LS}$ is equal to its variance, $v = \sigma^2/\Sigma z_i^2$ (from formula (3.4)), since its bias is 0. The mean squared error of $c\hat{\gamma}_{LS}$ is

$$\mathrm{MSE}(c\hat{\gamma}_{LS}) = c^2 v + (c-1)^2\gamma^2 \tag{8.3}$$

By taking the derivative with respect to c, setting it equal to 0, and solving for c, one finds that the smallest mean squared error is obtained for $c = \gamma^2/(\gamma^2 + v)$. In this expression for c, replace the unknown parameters γ and σ by the estimates $\hat{\gamma}_{LS}$ and $\hat{\sigma}_{LS}$ to obtain $\hat{c}$. By doing a little algebra, we find that $\hat{c} = \Sigma z_i^2/(\Sigma z_i^2 + \hat{\sigma}_{LS}^2/\hat{\gamma}_{LS}^2)$ and that $\hat{c}\hat{\gamma}_{LS}$ coincides with the ridge estimate (8.1).

Formula (8.3) is valid only if c is a nonrandom number. Since we have chosen a $\hat{c}$ that is random, involving the random quantities $\hat{\gamma}_{LS}$ and $\hat{\sigma}_{LS}$, the mean squared error of $\hat{\gamma} = \hat{c}\hat{\gamma}_{LS}$ is difficult to determine and, in fact, depends on the shape of the distribution of the random errors. Nevertheless, the argument given in the preceding paragraph at least suggests that $\hat{\gamma}$ might be more accurate than $\hat{\gamma}_{LS}$. However, it has been found that $\hat{\gamma}$ is more accurate than $\hat{\gamma}_{LS}$ only if γ is near 0. For simple regression, ridge estimation is not recommended. But to complete this section, we calculate the ridge estimates for the height data.

The Height Data. In a least-squares analysis of the data we find $\hat{\mu}_{LS} = \bar{y} = 166.8$, $\hat{\gamma}_{LS} = 5.093$, and $\hat{\sigma}_{LS} = 3.891$. Also, $\Sigma z_i^2 = 31.00$. The ridge estimates of μ and γ are $\hat{\mu} = 166.8$ and $\hat{\gamma} = [\Sigma z_i^2/(\Sigma z_i^2 + \hat{\sigma}_{LS}^2/\hat{\gamma}_{LS}^2)]\hat{\gamma}_{LS} = [31.00/(31.00 + 3.891^2/5.093^2)]5.093 = 4.999$. The estimated regression line is $\hat{Y} = 166.8 + 4.999Z$. We could put this in terms of the original variable X by substituting $Z = (X - 135.275)/5.687$ to obtain $\hat{Y} = 47.89 + 0.8790X$.

8.4 AN EXAMPLE OF MULTIPLE REGRESSION

The data in Table 8.2 describe 13 batches of cement. Each batch was made with four main ingredients: tricalcium aluminate (X_1), tricalcium silicate

Table 8.2 Cement Data

Identification Number of Batch	Heat (Y)	Ingredient 1 (X_1)	Ingredient 2 (X_2)	Ingredient 3 (X_3)	Ingredient 4 (X_4)
1	78.5	7	26	6	60
2	74.3	1	29	15	52
3	104.3	11	56	8	20
4	87.6	11	31	8	47
5	95.9	7	52	6	33
6	109.2	11	55	9	22
7	102.7	3	71	17	6
8	72.5	1	31	22	44
9	93.1	2	54	18	22
10	115.9	21	47	4	26
11	83.8	1	40	23	34
12	113.3	11	66	9	12
13	109.4	10	68	8	12

Source: Hald (1952, p. 647).

(X_2), tetracalcium alumino ferrite (X_3), and dicalcium silicate (X_4). The table shows the percentages of these ingredients used in each batch. The purpose of the experiment was to see how the relative amounts of these ingredients affect the heat (Y) evolved during the hardening of the cement. The heat is recorded in calories per gram of cement.

8.5 STANDARDIZATION

The explanatory variables are usually standardized before doing ridge estimation. In calculating the ridge estimate, all the explanatory variables are treated the same. (To be specific, in the notation of the next section, the same number k is added to each diagonal entry of $\mathbf{Z}'\mathbf{Z}$.) Therefore it is sensible to standardize the variables so that their units of measurement are comparable.

Some statisticians also recommend standardization before doing least-squares regression. This can be beneficial in three ways. First, it reduces round-off error in the inversion of the matrix $\mathbf{X}'\mathbf{X}$. Second, it can increase the interpretability of the variables and their regression coefficients. For example, consider the variable weight for a group of 9-year-old girls. To say that the weight of one of the girls is 24 kilograms may not be as meaningful to you as to say that she is 1.47 standard deviations below the average weight of the group.

Third, standardization makes it possible to directly compare the regression coefficients of different variables. Suppose $\hat{Y} = -171 + 1.92X_1 + 0.286X_2$ is

Table 8.3 Standardized Cement Data

Identification Number of Batch	Heat (Y)	Ingredient 1 (Standardized) (Z_1)	Ingredient 2 (Standardized) (Z_2)	Ingredient 3 (Standardized) (Z_3)	Ingredient 4 (Standardized) (Z_4)
1	78.5	−0.0785	−1.4237	−0.9007	1.7923
2	74.3	−1.0985	−1.2309	0.5044	1.3144
3	104.3	0.6015	0.5042	−0.5885	−0.5974
4	87.6	0.6015	−1.1024	−0.5885	1.0156
5	95.9	−0.0785	0.2472	−0.9007	0.1792
6	109.2	0.6015	0.4400	−0.4323	−0.4779
7	102.7	−0.7585	1.4682	0.8167	−1.4338
8	72.5	−1.0985	−1.1024	1.5973	0.8364
9	93.1	−0.9285	0.3757	0.9728	−0.4779
10	115.9	2.3015	−0.0742	−1.2130	−0.2390
11	83.8	−1.0985	−0.5240	1.7534	0.2390
12	113.3	0.6015	1.1469	−0.4323	−1.0754
13	109.4	0.4315	1.2754	−0.5885	−1.0754

an estimated regression equation with Y = lung capacity in centiliters, X_1 = height in centimeters, and X_2 = weight in kilograms. It does not make sense to compare the coefficients 1.92 and 0.286. Now standardize the explanatory variables. Call the standardized variables Z_1 and Z_2 and suppose the estimated regression equation becomes $\hat{Y} = 193 + 12.9Z_1 + 3.28Z_2$. Now we can compare 12.9 and 3.28 to conclude that differences in lung capacity are due more to differences in height than to differences in weight.

The standardized cement data are shown in Table 8.3.

8.6 ESTIMATING THE REGRESSION COEFFICIENTS

First we must write down the model. In terms of the original unstandardized variables, the model is

$$y_i = \beta_0 + \beta_1 x_{i1} + \beta_2 x_{i2} + \beta_3 x_{i3} + \beta_4 x_{i4} + e_i$$

In terms of the standardized variables, let us use the notation

$$y_i = \mu + \gamma_1 z_{i1} + \gamma_2 z_{i2} + \gamma_3 z_{i3} + \gamma_4 z_{i4} + e_i$$

In matrix notation,

$$\boldsymbol{y} = \mathbf{1}\mu + \boldsymbol{Z\gamma} + \boldsymbol{e}$$

where

$$\mathbf{y} = \begin{bmatrix} y_1 \\ y_2 \\ \vdots \\ y_n \end{bmatrix}, \quad \mathbf{1} = \begin{bmatrix} 1 \\ 1 \\ \vdots \\ 1 \end{bmatrix}, \quad \mathbf{Z} = \begin{bmatrix} z_{11} & z_{12} & z_{13} & z_{14} \\ z_{21} & z_{22} & z_{23} & z_{24} \\ \vdots & \vdots & \vdots & \vdots \\ z_{n1} & z_{n2} & z_{n3} & z_{n4} \end{bmatrix},$$

$$\boldsymbol{\gamma} = \begin{bmatrix} \gamma_1 \\ \gamma_2 \\ \gamma_3 \\ \gamma_4 \end{bmatrix}, \quad \mathbf{e} = \begin{bmatrix} e_1 \\ e_2 \\ \vdots \\ e_n \end{bmatrix}$$

and $n = 13$.

The Ridge Estimates. The least-squares estimates of μ and $\boldsymbol{\gamma}$ are

$$\hat{\mu}_{\text{LS}} = \bar{y}$$

$$\hat{\boldsymbol{\gamma}}_{\text{LS}} = (\mathbf{Z}'\mathbf{Z})^{-1}\mathbf{Z}'\mathbf{y}$$

For the ridge estimates, we keep $\hat{\mu} = \bar{y}$ and let

$$\hat{\boldsymbol{\gamma}} = (\mathbf{Z}'\mathbf{Z} + k\mathbf{I})^{-1}\mathbf{Z}'\mathbf{y} \tag{8.4}$$

where $k = 4\hat{\sigma}^2_{\text{LS}}/\|\hat{\boldsymbol{\gamma}}_{\text{LS}}\|^2$, $\hat{\sigma}^2_{\text{LS}}$ is the least-squares estimate of σ^2 in formula (3.12) with $n = 13$ and $p = 4$, and $\mathbf{I}$ is the identity matrix,

$$\mathbf{I} = \begin{bmatrix} 1 & 0 & 0 & 0 \\ 0 & 1 & 0 & 0 \\ 0 & 0 & 1 & 0 \\ 0 & 0 & 0 & 1 \end{bmatrix}$$

(For a vector $\mathbf{v}$, the notation $\|\mathbf{v}\|$ denotes the length of the vector; that is, $\|\mathbf{v}\| = \sqrt{\Sigma v_i^2} = \sqrt{\mathbf{v}'\mathbf{v}}$.)

In general, $\mathbf{Z}$ is an $n \times p$ matrix, $\mathbf{I}$ is the $p \times p$ identity matrix, and $k = p\hat{\sigma}^2_{\text{LS}}/\|\hat{\boldsymbol{\gamma}}_{\text{LS}}\|^2$.

A Partial Justification of (8.4). The justification given in Section 8.3 for simple ridge regression can be extended to multiple ridge regression. The argument is not completely precise, as is pointed out below. Further justification is given in the next section.

Note that when $p = 1$, formula (8.4) is the same as (8.1). Thus (8.4) can be viewed as a generalization of (8.1). As was said in the justification of the latter formula, we can again say that ridge estimation tries to improve the accuracy of the least-squares estimate by shrinking it. To see this, write $\hat{\gamma} = C\hat{\gamma}_{LS}$, where $C = (Z'Z + kI)^{-1}Z'Z$ and check that $\|C\hat{\gamma}_{LS}\| < \|\hat{\gamma}_{LS}\|$. The accuracy of the vector of estimates $\hat{\gamma}$ can be measured by its *total mean squared error*, that is, the sum of the mean squared errors of its components, $\Sigma\text{MSE}(\hat{\gamma}_j)$. If one expresses the total mean squared error as a function of k, takes its derivative, and sets this equal to 0, one finds that the smallest total mean squared error is obtained for k satisfying $k = \sigma^2\,\text{trace}(B)/\gamma' B\gamma$, where B is a matrix which, unfortunately, involves k. (The notation trace(B) is used for the sum of the diagonal entries of B.) So there is no explicit expression for the best value of k. We attempt to get an explicit value of k that is good, though not necessarily best, by replacing B by the identity matrix I. (There is a gap in this line of argument because we can provide no convincing reason for choosing I.) Now $k = p\sigma^2/\gamma'\gamma$. Substituting least-squares estimates for σ^2 and γ, we obtain (8.4).

The Cement Data. From the standardized data in Table 8.3 we first calculate the least-squares estimates:

$$\hat{\mu}_{LS} = 95.42, \quad \hat{\gamma}_{LS} = \begin{bmatrix} 9.124 \\ 7.937 \\ 0.653 \\ -2.412 \end{bmatrix}, \quad \text{and} \quad \hat{\sigma}_{LS} = 2.466$$

Next calculate $\|\hat{\gamma}_{LS}\|^2 = (9.124)^2 + (7.937)^2 + (0.653)^2 + (-2.412)^2 = 152.5$ and $k = 4(2.446)^2/152.5 = 0.1569$. Now calculate formula (8.4):

$$\hat{\gamma} = \begin{bmatrix} 7.644 \\ 4.667 \\ -0.910 \\ -5.835 \end{bmatrix}$$

The estimated regression equation is $\hat{Y} = 95.42 + 7.644Z_1 + 4.667Z_2 - 0.910Z_3 - 5.835Z_4$.

8.7 COLLINEARITY

Ridge estimation can be justified in terms of the problem of collinearity, which is what originally motivated Hoerl to invent the ridge method. A set of variables are exactly *collinear* if one of them is a linear function of the others. They are approximately collinear if one of them is approximately a linear function of the others. We use "collinear" to mean "approximately collinear".

Table 8.4 Sample Correlations Between the Explanatory Variables in the Cement Data

Pair of Variables	Sample Correlation
X_1, X_2	0.229
X_1, X_3	-0.824
X_1, X_4	-0.245
X_2, X_3	-0.139
X_2, X_4	-0.973
X_3, X_4	0.030

Two variables are collinear if they are highly correlated. The sample correlations between the explanatory variables in the cement data are shown in Table 8.4. There is a very high negative correlation of -0.973 between X_2 and X_4. Looking at the data we can see the reason for this. Except for batches 1 and 5, the total percentage of silicon compounds, $X_2 + X_4$, has been kept close to 77, so that X_4 is approximately equal to $77 - X_2$. This makes it difficult to distinguish between the effects of X_2 and X_4. For instance, the four highest values of X_4, which are 60, 52, 47, and 44, occur for values of heat that are below the average heat of 95.4. So it looks like high amounts of ingredient 4 cause less heat to be evolved. But, corresponding to the high negative correlation between X_2 and X_4, we see that the four highest values of X_4 occur with the four lowest values of X_2. So maybe ingredient 4 does not have much effect on heat and it is really the low amounts of ingredient 2 that cause less heat to be evolved. Thus we see how collinearity of two explanatory variables makes it difficult to pin down the effect of either variable, which implies poor accuracy for the estimates of their coefficients.

There is also a high negative correlation of -0.824 between X_1 and X_3, which arises because the total percentage of aluminum compounds, $X_1 + X_3$, has been kept close to 20.

Variance Inflation Factors. A collinear relationship involving three or more variables will not necessarily be revealed by the pairwise correlations. A better indication of whether there is a problem of collinearity is given by the variance inflation factors (VIFs). The *variance inflation factor* of an explanatory variable X_j is a measure of how closely X_j is related to the other explanatory variables (but only considering linear relationships). Specifically,

$$\text{VIF}_j = \frac{1}{1 - R_j^2}$$

where R_j^2 is the coefficient of determination when X_j is regressed on the other explanatory variables. If X_j is exactly collinear with the other variables, that is, if X_j is an exact linear function of them, then $R_j^2 = 1$ and $\text{VIF}_j = \infty$. If X_j is completely uncorrelated with the other variables, then $R_j^2 = 0$ and $\text{VIF}_j = 1$.

For the cement data, the VIFs for X_1, X_2, X_3, and X_4 are 38.5, 254.4, 46.9, and 282.5. There is a collinearity problem for a variable X_j if R_j^2 is close to 1, that is, if VIF_j is large. The computer package Minitab prints a warning when VIF_j is larger than 100.

There is a direct connection between the variance inflation factor of X_j and the standard deviation of the least-squares estimate of its regression coefficient:

$$\text{SD}(\hat{\beta}_j) = \sqrt{\text{VIF}_j}\left(\frac{\sigma}{s_j}\right)\Big/\sqrt{n-1}$$

where $\hat{\beta}_j = \hat{\beta}_{\text{LS}j}$ and s_j is the standard deviation of the observed values of X_j. Hence the size of $\text{SD}(\hat{\beta}_j)$ is determined by three factors: a factor $\sqrt{\text{VIF}_j}$ due to the relationship of X_j with the other explanatory variables, a factor σ/s_j depending on the variation of the random errors relative to the variation of the measurements of X_j, and a factor $1/\sqrt{n-1}$ depending on the sample size. Inaccuracy of $\hat{\beta}_j$ can be due to collinearity of X_j with other explanatory variables or to large random errors or to a small sample size. The phrase "variance inflation factor" comes from the fact that $\text{Var}(\hat{\beta}_j)$ is VIF_j times larger than what it would be if X_j were uncorrelated with the other explanatory variables.

Matrix Formulation of the Problem of Collinearity. In terms of the matrix $\mathbf{Z}$ of explanatory variables, collinearity means that some column of $\mathbf{Z}$ is approximately a linear combination of the other columns. This implies that the matrix $\mathbf{Z}'\mathbf{Z}$, which must be inverted to calculate the least-squares estimate of $\boldsymbol{\gamma}$, is nearly singular. Inverting a singular matrix is like taking the reciprocal of the number 0; it is not a valid operation. Inverting a nearly singular matrix is similar to taking the reciprocal of a very small number; some of the entries in the inverse matrix are likely to be very large.

The variance of the least-squares estimate $\hat{\gamma}_{\text{LS}j}$ is equal to σ^2 times the jth diagonal entry in $(\mathbf{Z}'\mathbf{Z})^{-1}$. Therefore near-singularity of $\mathbf{Z}'\mathbf{Z}$ is likely to be associated with large variances for some of the least-squares estimates. This indicates how collinearity leads to inaccuracy of regression estimates.

Besides the statistical problem of large variances for the estimates, collinearity also poses a computational problem. It is difficult to achieve numerical accuracy in computing the inverse of a nearly singular matrix.

Further Justification of Ridge Estimation. By viewing the problem of collinearity as the problem of the near-singularity of $\mathbf{Z}'\mathbf{Z}$, we are led to the method of ridge regression, which modifies $\mathbf{Z}'\mathbf{Z}$ so that it is farther from singularity. The matrix $\mathbf{Z}'\mathbf{Z}$ is modified so that it is closer to what it would be for data in which there is no collinearity, that is, data in which all the explanatory variables are uncorrelated with one another. We now need to know what $\mathbf{Z}'\mathbf{Z}$ looks like for such data.

The matrix $\mathbf{Z}'\mathbf{Z}$ is $n - 1$ times the sample correlation matrix of the explanatory variables. To verify this, note that the (j, k) entry of $\mathbf{Z}'\mathbf{Z}$ is $\Sigma_i z_{ij} z_{ik}$. Since $z_{ij} = (x_{ij} - \bar{x}_j)/s_j$, where $\bar{x}_j$ and s_j are the sample mean and standard deviation of the observed values of variable X_j, the (j, k) entry of $\mathbf{Z}'\mathbf{Z}$ is $\Sigma_i (x_{ij} - \bar{x}_j)(x_{ik} - \bar{x}_k)/(s_j s_k)$. Recall that $s_{jk} = \Sigma_i (x_{ij} - \bar{x}_j)(x_{ik} - \bar{x}_k)/(n - 1)$ is the sample covariance of X_j and X_k. So the (j, k) entry is $(n - 1)s_{jk}/(s_j s_k)$. By definition, $s_{jk}/(s_j s_k)$ is the sample correlation between X_j and X_k.

In the most favorable case, in which all the explanatory variables are uncorrelated, the sample correlation matrix is simply the identity matrix $\mathbf{I}$, and so $\mathbf{Z}'\mathbf{Z} = (n - 1)\mathbf{I}$. When there is collinearity, we can move $\mathbf{Z}'\mathbf{Z}$ closer to the most favorable case by adding a multiple of $\mathbf{I}$ to $\mathbf{Z}'\mathbf{Z}$, that is, by replacing $\mathbf{Z}'\mathbf{Z}$ by $\mathbf{Z}'\mathbf{Z} + k\mathbf{I}$. This leads to (8.4).

NOTES

8.3a. The formula for $\hat{\gamma}_{\text{LS}}$ is obtained from (3.2). Direct application of (3.2) gives us $\hat{\gamma}_{\text{LS}} = \Sigma(z_i - \bar{z})(y_i - \bar{y})/\Sigma(z_i - \bar{z})^2$. Since $\bar{z} = 0$, this becomes $\hat{\gamma}_{\text{LS}} = \Sigma z_i(y_i - \bar{y})/\Sigma z_i^2$. Moreover, $\Sigma z_i(y_i - \bar{y}) = \Sigma z_i y_i$ because $\Sigma z_i \bar{y} = \bar{y}\Sigma z_i = \bar{y}(n\bar{z}) = 0$.

8.3b. To verify (8.2), let $\xi = E(\hat{\gamma})$ and note that $\text{Var}(\hat{\gamma}) = E[(\hat{\gamma} - \xi)^2]$. Now $\text{MSE}(\hat{\gamma}) = E[(\hat{\gamma} - \gamma)^2] = E[((\hat{\gamma} - \xi) + (\xi - \gamma))^2] = E[(\hat{\gamma} - \xi)^2 + (\xi - \gamma)^2 + 2(\hat{\gamma} - \xi)(\xi - \gamma)] = E[(\hat{\gamma} - \xi)^2] + (\xi - \gamma)^2 + 2(\xi - \gamma)E(\hat{\gamma} - \xi) = \text{Var}(\hat{\gamma}) + [E(\hat{\gamma}) - \gamma]^2$ because $E(\hat{\gamma} - \xi) = E(\hat{\gamma}) - \xi = 0$.

8.3c. Formula (8.3) follows from (8.2). By (8.2), $\text{MSE}(c\hat{\gamma}_{\text{LS}}) = \text{Var}(c\hat{\gamma}_{\text{LS}}) + [E(c\hat{\gamma}_{\text{LS}}) - \gamma]^2$, which equals $c^2\text{Var}(\hat{\gamma}_{\text{LS}}) + [cE(\hat{\gamma}_{\text{LS}}) - \gamma]^2 = c^2 v + [c\gamma - \gamma]^2 = c^2 v + (c - 1)^2\gamma^2$. The derivative with respect to c is $2cv + 2(c - 1)\gamma^2 = 2[c(v + \gamma^2) - \gamma^2]$, which is 0 when $c(v + \gamma^2) = \gamma^2$.

8.3d. It is difficult to figure out the exact mean squared error of the ridge estimate $\hat{\gamma}$ but we can estimate it by computer simulations. For example, let us simulate a model similar to the process that generated the height data. Consider the model $y_i = 170 + 5z_i + e_i$ for $i = 1, 2, \ldots, 32$, where the z_i's are the standardized heights in Table 8.1. Suppose the random errors e_i are normally distributed with mean 0 and standard deviation 4.

Using a pseudo-random number generator, we can generate 32 pseudo-independent random numbers e_i from a normal distribution with mean 0 and standard deviation 4. Add each e_i to $170 + 5z_i$ to obtain y_i and then apply (8.1) to obtain the ridge estimate $\hat{\gamma}$. We know the true value of the parameter γ is 5, and so the accuracy of $\hat{\gamma}$ can be seen from the difference $\hat{\gamma} - 5$. Repeat this a large number of times, say, 500. Each time, a sample of 32 y_i's is generated and $\hat{\gamma}$ is calculated. Thus we obtain 500 values of $\hat{\gamma}$. The average of the 500 values of $(\hat{\gamma} - 5)^2$ is a good estimate of $\text{MSE}(\hat{\gamma})$.

In such a simulation, $\text{MSE}(\hat{\gamma})$ was estimated to be 0.541. The value of $\text{MSE}(\hat{\gamma}_{\text{LS}})$ is known to be $\text{Var}(\hat{\gamma}_{\text{LS}}) = \sigma^2/\Sigma z_i^2 = 4^2/31 = 0.516$. So the ridge estimate has larger mean squared error than the least-squares estimate by about 5% $(= (0.541 - 0.516)/0.516)$.

By running simulations for other values of γ, it was found that $\text{MSE}(\hat{\gamma}) < \text{MSE}(\hat{\gamma}_{\text{LS}})$ only when $-1 < \gamma < 1$.

8.6a. The formulas for $\hat{\mu}_{\text{LS}}$ and $\hat{\boldsymbol{\gamma}}_{\text{LS}}$ can be verified as follows. In the model $\boldsymbol{y} = \boldsymbol{X\beta} + \boldsymbol{e}$ the formula for the vector of least-squares regression estimates is $\boldsymbol{\beta}_{\text{LS}} = (\boldsymbol{X}'\boldsymbol{X})^{-1}\boldsymbol{X}'\boldsymbol{y}$. To apply this to the standardized model $\boldsymbol{y} = \mathbf{1}\mu + \boldsymbol{Z\gamma} + \boldsymbol{e}$, replace $\boldsymbol{X}$ by $[\mathbf{1} \quad \boldsymbol{Z}]$ and $\boldsymbol{\beta}$ by $\begin{bmatrix} \mu \\ \gamma \end{bmatrix}$. Then $\boldsymbol{X}'\boldsymbol{X}$ is replaced by

$$[\mathbf{1} \quad \boldsymbol{Z}]'[\mathbf{1} \quad \boldsymbol{Z}] = \begin{bmatrix} \mathbf{1}' \\ \boldsymbol{Z}' \end{bmatrix}[\mathbf{1} \quad \boldsymbol{Z}] = \begin{bmatrix} \mathbf{1}'\mathbf{1} & \mathbf{1}'\boldsymbol{Z} \\ \boldsymbol{Z}'\mathbf{1} & \boldsymbol{Z}'\boldsymbol{Z} \end{bmatrix} = \begin{bmatrix} n & 0 \\ 0 & \boldsymbol{Z}'\boldsymbol{Z} \end{bmatrix}$$

Note that $\mathbf{1}'\boldsymbol{Z} = \mathbf{0}$ because, being standardized, each column of $\boldsymbol{Z}$ sums to 0. And $\boldsymbol{X}'\boldsymbol{y}$ is replaced by

$$[\mathbf{1} \quad \boldsymbol{Z}]'\boldsymbol{y} = \begin{bmatrix} \mathbf{1}' \\ \boldsymbol{Z}' \end{bmatrix}\boldsymbol{y} = \begin{bmatrix} \mathbf{1}'\boldsymbol{y} \\ \boldsymbol{Z}'\boldsymbol{y} \end{bmatrix} = \begin{bmatrix} \Sigma y_i \\ \boldsymbol{Z}'\boldsymbol{y} \end{bmatrix}$$

Hence

$$\begin{bmatrix} \hat{\mu}_{\text{LS}} \\ \hat{\boldsymbol{\gamma}}_{\text{LS}} \end{bmatrix} = \begin{bmatrix} n & 0 \\ 0 & \boldsymbol{Z}'\boldsymbol{Z} \end{bmatrix}^{-1}\begin{bmatrix} \Sigma y_i \\ \boldsymbol{Z}'\boldsymbol{y} \end{bmatrix} = \begin{bmatrix} \bar{y} \\ (\boldsymbol{Z}'\boldsymbol{Z})^{-1}\boldsymbol{Z}'\boldsymbol{y} \end{bmatrix}$$

8.6b. The formula $k = p\hat{\sigma}_{\text{LS}}^2/\|\hat{\boldsymbol{\gamma}}_{\text{LS}}\|^2$ was suggested by Hoerl, Kennard, and Baldwin (1975). Many other formulas for k have been proposed; see Draper and Van Nostrand (1979) and Van Nostrand (1980).

8.6c. Let $\text{TMSE}(\hat{\boldsymbol{\gamma}})$ be the total mean squared error of $\hat{\boldsymbol{\gamma}}$. Formula (8.2) generalizes to

$$\text{TMSE}(\hat{\boldsymbol{\gamma}}) = \text{trace}[\text{Cov}(\hat{\boldsymbol{\gamma}})] + \|E(\hat{\boldsymbol{\gamma}}) - \boldsymbol{\gamma}\|^2$$

The ridge estimate $\hat{\boldsymbol{\gamma}}$ has the form $\hat{\boldsymbol{\gamma}} = \boldsymbol{A}\boldsymbol{y}$, where $\boldsymbol{A} = (\boldsymbol{Z}'\boldsymbol{Z} + k\boldsymbol{I})^{-1}\boldsymbol{Z}'$. Regard k as a nonrandom constant. Then the entries of $\boldsymbol{A}$ are nonrandom constants, so we can use rules (a) and (b) in Note 3.7 to calculate $E(\hat{\boldsymbol{\gamma}})$ and $\text{Cov}(\hat{\boldsymbol{\gamma}})$, which can be put into the formula above to obtain $\text{TMSE}(\hat{\boldsymbol{\gamma}})$. Taking its derivative with respect to k is somewhat involved and will not be presented here. Rules for differentiating matrices can be found in Searle (1982). The derivative turns out to be $-2\sigma^2 \,\text{trace}(\boldsymbol{B}) + 2k\boldsymbol{\gamma}'\boldsymbol{B}\boldsymbol{\gamma}$, where $\boldsymbol{B} = \boldsymbol{Z}'\boldsymbol{Z}(\boldsymbol{Z}'\boldsymbol{Z} + k\boldsymbol{I})^{-3}$. The derivative is 0 when $k = \sigma^2 \,\text{trace}(\boldsymbol{B})/\boldsymbol{\gamma}'\boldsymbol{B}\boldsymbol{\gamma}$.

8.6d. There always exists a value of k for which the TMSE of the ridge estimate is smaller than the TMSE of the least-squares estimate. To see this we use the expression given in Note 8.6c for the derivative of $\text{TMSE}(\hat{\boldsymbol{\gamma}})$ with respect to k. The derivative at $k = 0$ is $-2\sigma^2\text{trace}(\boldsymbol{B}_0)$, where $\boldsymbol{B}_0 = (\boldsymbol{Z}'\boldsymbol{Z})^{-2}$. Since the trace of $(\boldsymbol{Z}'\boldsymbol{Z})^{-2}$ is positive, the derivative at $k = 0$ is negative, so $\text{TMSE}(\hat{\boldsymbol{\gamma}})$ is decreasing. Note that at $k = 0$, $\hat{\boldsymbol{\gamma}} = \hat{\boldsymbol{\gamma}}_{\text{LS}}$. Therefore, for small positive values of k, $\text{TMSE}(\hat{\boldsymbol{\gamma}}) < \text{TMSE}(\hat{\boldsymbol{\gamma}}_{\text{LS}})$.

8.6e. Another justification of (8.4) can be given by taking a Bayesian viewpoint in which $\boldsymbol{\gamma}$ is regarded as a random vector. Conditional on a fixed value of $\boldsymbol{\gamma}$, we suppose $\boldsymbol{y}$ has a multivariate normal distribution with mean vector $\mathbf{1}\mu + \boldsymbol{Z}\boldsymbol{\gamma}$ and covariance matrix $\sigma^2\boldsymbol{I}$. And we suppose $\boldsymbol{\gamma}$ is a random vector having a multivariate normal distribution with mean vector $\mathbf{0}$ and covariance matrix $\tau^2\boldsymbol{I}$. In the Bayesian approach to statistical analysis (see Chapter 7) the prior distribution of $\boldsymbol{\gamma}$ is combined with the data to produce the posterior distribution of $\boldsymbol{\gamma}$, that is, its conditional distribution given $\boldsymbol{y}$. It can be shown (see the lemma on p. 4 of Lindley and Smith, 1972) that the posterior mean vector of $\boldsymbol{\gamma}$ is $(\boldsymbol{Z}'\boldsymbol{Z} + \kappa\boldsymbol{I})^{-1}\boldsymbol{Z}'\boldsymbol{y}$, where $\kappa = \sigma^2/\tau^2$. We can show that $k = p\hat{\sigma}^2_{\text{LS}}/\|\hat{\boldsymbol{\gamma}}_{\text{LS}}\|^2$ is a reasonable estimate of κ by showing that $\|\hat{\boldsymbol{\gamma}}_{\text{LS}}\|^2/p$ is a reasonable estimate of τ^2. To see this, recall that, based on the distribution of $\boldsymbol{y}$ conditional on $\boldsymbol{\gamma}$, a reasonable estimate of σ^2 is $\hat{\sigma}^2_{\text{LS}} = \|\boldsymbol{y} - (\mathbf{1}\bar{y} + \boldsymbol{Z}\hat{\boldsymbol{\gamma}}_{\text{LS}})\|^2/(n - (p + 1))$. Similarly, based on the distribution of $\boldsymbol{\gamma}$, a reasonable estimate of τ^2 is $\|\boldsymbol{\gamma} - \mathbf{0}\|^2/(p - 0)$.

8.6f. To estimate the total mean squared error (TMSE) of the ridge estimate $\hat{\boldsymbol{\gamma}}$, we can use computer simulations. For example, let us simulate a model similar to the process that generated the cement data. Consider the model $y_i = 95 + 7.6z_{i1} + 4.7z_{i2} - 0.9z_{i3} - 5.8z_{i4} + e_i$ for $i = 1, 2, \ldots, 13$, where the z_{ij}'s are the standardized variables in Table 8.3. Suppose the random errors e_i are normally distributed with mean 0 and standard deviation 2.4.

Using a pseudo-random number generator, we can generate 13 pseudo-independent random numbers e_i from a normal distribution with mean 0 and standard deviation 2.4. Add each e_i to $95 + 7.6z_{i1} + 4.7z_{i2} - 0.9z_{i3} - 5.8z_{i4}$ to obtain y_i and then apply (8.4) to obtain the vector of ridge estimates

$\hat{\gamma} = (\hat{\gamma}_1, \hat{\gamma}_2, \hat{\gamma}_3, \hat{\gamma}_4)$. We know the true value of the parameter vector γ is $(7.6, 4.7, -0.9, -5.8)$, and so the accuracy of $\hat{\gamma}$ can be seen from the vector of differences $(\hat{\gamma}_1 - 7.6,\ \hat{\gamma}_2 - 4.7,\ \hat{\gamma}_3 + 0.9, \hat{\gamma}_4 + 5.8)$. Repeat this a large number of times, say, 500. Each time, a sample of 13 y_i's is generated and $\hat{\gamma}$ is calculated. Thus we obtain 500 vectors $\hat{\gamma}$. The average of the 500 values of $(\hat{\gamma}_1 - 7.6)^2 + (\hat{\gamma}_2 - 4.7)^2 + (\hat{\gamma}_3 + 0.9)^2 + (\hat{\gamma}_4 + 5.8)^2$ is a good estimate of $\text{TMSE}(\hat{\gamma})$.

In such a simulation, $\text{TMSE}(\hat{\gamma})$ was estimated to be 77. Using Note 8.6c, the value of $\text{TMSE}(\hat{\gamma}_{\text{LS}})$ can be calculated to be $\text{trace}[\text{Cov}(\hat{\gamma}_{\text{LS}})] = \sigma^2\,\text{trace}[(Z'Z)^{-1}] = (2.4)^2\,\text{trace}[(Z'Z)^{-1}] = 299$. So the ridge estimates have smaller total mean squared error than the least-squares estimates by about 74% $(= (299 - 77)/299)$.

8.7a. For an example in which collinearity is not revealed by the pairwise correlations, consider the following data.

X_1	X_2	X_3
1	2	3
1	3	2
2	1	3
2	3	1
3	1	2
3	2	1

The three pairwise correlations are all -0.50, which is not high. But the three variables are exactly collinear because $X_1 = 6 - X_2 - X_3$.

8.7b. Let us show $\text{SD}(\hat{\beta}_j) = \sqrt{\text{VIF}_j}\,(\sigma/s_j)/\sqrt{n-1}$, or equivalently, $\text{Var}(\hat{\beta}_j) = (\text{VIF}_j)(\sigma^2/s_j^2)/(n-1)$. By reindexing we can suppose $j = p$. Recall from Section 3.10 that $\text{Var}(\hat{\beta}_p)$ is σ^2 times the last diagonal entry in $(X'X)^{-1}$. Let x denote the last column of X, containing the observed values of variable X_p, and let W denote the matrix consisting of the other columns. Thus $X = [W \quad x]$. Then

$$X'X = \begin{bmatrix} W'W & W'x \\ x'W & x'x \end{bmatrix}$$

The last diagonal entry in $(X'X)^{-1}$ is $1/(x'x - x'W(W'W)^{-1}W'x)$. This is not hard to verify for the case $p = 1$ because then W is simply a vector w (in fact, a vector of 1's) and $X'X$ is a 2×2 matrix, which is easy to invert.

Note that $x'x - x'W(W'W)^{-1}W'x = \text{SSR}_p$, the sum of squared residuals for a least-squares regression analysis of X_p as a linear function of $X_1, \ldots, X_{p-1}$. We must show $\sigma^2/\text{SSR}_p = \sigma^2\text{VIF}_p/((n-1)s_p^2)$. Since

$\text{VIF}_p = 1/(1 - R_p^2)$ and $(n-1)s_p^2 = \Sigma(x_{ip} - \bar{x}_p)^2$, we must show $\text{SSR}_p = (1 - R_p^2)\Sigma(x_{ip} - \bar{x}_p)^2$, or $R_p^2 = 1 - \text{SSR}_p/\Sigma(x_{ip} - \bar{x}_p)^2$. This is shown in Note 3.11a.

8.7c. The variance inflation factor VIF_j can be calculated as $n - 1$ times the jth diagonal entry of $(\mathbf{Z}'\mathbf{Z})^{-1}$. As for why this works, note that the columns of $\mathbf{Z}$ are standardized and so the standard deviation of the observed values of Z_j is 1. Hence, by the second formula in Section 8.7, $\text{Var}(\hat{\gamma}_{\text{LS}j}) = \text{VIF}_j\sigma^2/(n-1)$. As in Note 3.10, it is also true that $\text{Var}(\hat{\gamma}_{\text{LS}j})$ is equal to σ^2 times the jth diagonal entry of $(\mathbf{Z}'\mathbf{Z})^{-1}$.

8.7d. The following informal argument shows how collinearity of the columns of $\mathbf{Z}$ implies near-singularity of $\mathbf{Z}'\mathbf{Z}$. We use the fact that exact singularity (that is, noninvertibility) of a square matrix is equivalent to exact collinearity of its columns. Let the columns of $\mathbf{Z}$ be $\mathbf{z}_1, \mathbf{z}_2, \ldots, \mathbf{z}_p$. Collinearity means that $a_1\mathbf{z}_1 + a_2\mathbf{z}_2 + \cdots + a_p\mathbf{z}_p \approx \mathbf{0}$ for numbers a_j not all 0. This can also be expressed as $\mathbf{Za} \approx \mathbf{0}$, where $\mathbf{a} = (a_1, a_2, \ldots, a_p)$. Then $\mathbf{Z}'\mathbf{Za} \approx \mathbf{0}$, so the columns of $\mathbf{Z}'\mathbf{Z}$ are collinear.

Additional Reading. To read more about ridge regression, see Section 6.7 in Draper and Smith (1981), Section 5.5 in Freund and Minton (1979), or Section 10.3 in Gunst and Mason (1980). Ridge regression was introduced into the statistical literature in two journal articles by Hoerl and Kennard (1970a, b). Marquardt and Snee (1975) have written an expository article.

REFERENCES

Draper, N. R., and H. Smith (1981). *Applied Regression Analysis*, 2nd ed. Wiley, New York.

Draper, N. R., and R. C. Van Nostrand (1979). Ridge regression and James-Stein estimation: review and comments. *Technometrics*, vol. 21, pp. 451–466.

Freund, R. J., and P. D. Minton (1979). *Regression Methods: A Tool for Data Analysis*. Marcel Dekker, New York.

Gunst, R. F., and R. L. Mason (1980). *Regression Analysis and Its Application: A Data-Oriented Approach*. Marcel Dekker, New York.

Hald, A. (1952). *Statistical Theory with Engineering Applications*. Wiley, New York.

Hoerl, A. E. (1962). Application of ridge analysis to regression problems. *Chemical Engineering Progress*, vol. 58, no. 3, pp. 54–59.

Hoerl, A. E., and R. W. Kennard (1970a). Ridge regression: biased estimation for nonorthogonal problems. *Technometrics*, vol. 12, pp. 55–67.

Hoerl, A. E., and R. W. Kennard (1970b). Ridge regression: applications to nonorthogonal problems. *Technometrics*, vol. 12, pp. 69–82.

Hoerl, A. E., R. W. Kennard, and K. R. Baldwin (1975). Ridge regression: some simulations. *Communications in Statistics A*, vol. 4, pp. 105–123.

Lindley, D. V., and A. F. M. Smith (1972). Bayes estimates for the linear model. *Journal of the Royal Statistical Society, Series B*, vol. 33, pp. 1–44.

Marquardt, D. W., and R. D. Snee (1975). Ridge regression in practice. *American Statistician*, vol. 29, pp. 3–20.

Searle, S. R. (1982). *Matrix Algebra Useful for Statistics*. Wiley, New York.

Van Nostrand, R. C. (1980). Comment on "A critique of some ridge regression methods" by G. Smith and F. Campbell. *Journal of the American Statistical Association*, vol. 75, pp. 92–94.

Weisberg, S. (1985). *Applied Linear Regression*, 2nd ed. Wiley, New York.

CHAPTER 9

Comparisons

Then to the rolling Heav'n itself I cried
Asking, "What Lamp had Destiny to guide
Her little Children stumbling in the Dark?"
And—"A blind Understanding!" Heav'n replied.

KHAYYAM NAISHAPURI: *Persian astronomer and poet (1048–1131)*

9.1 INTRODUCTION

The preceding six chapters present six different methods for obtaining estimates and test statistics for the regression coefficients in a linear regression model. In this chapter we compare these methods with one another by considering what statistical theory says about their properties and by looking at their performances on some real data sets.

9.2 COMPARISON OF PROPERTIES

Let us consider the properties of the six methods. These properties are described briefly, without getting into all the mathematical details that would be required to fully explain some of them. References to more details are given in the Notes.

Some Terminology. A property is called an *optimal* property if it states that a procedure is best, where "best" is judged according to a particular criterion and attention may be restricted to a particular class of procedures.

A property is called *asymptotic* if it concerns the limiting behavior of a procedure as the size of the sample becomes infinitely large. Asymptotic properties are also called *large-sample* properties. Since real data occur in samples of finite size, asymptotic properties are not as directly relevant as finite-sample properties, but asymptotic results are often more readily obtainable.

Most estimates, if they are at all reasonable, have the asymptotic property of consistency. Let $\hat{\theta}$ be an estimate of a parameter θ. We say $\hat{\theta}$ is *consistent* if, when the sample size is very large, $\hat{\theta}$ is very close to θ with high probability. Many estimates are also asymptotically normal. We say $\hat{\theta}$ is *consistent and asymptotically normal* (CAN) if, when the sample size is very large, the probability distribution of $\hat{\theta}$ is approximately normal with mean θ and variance of the form τ^2/n, where τ^2 does not depend on n. We call τ^2 the *asymptotic variance* of $\hat{\theta}$. The large-sample performances of two CAN estimates of the same parameter can be compared by comparing their asymptotic variances. Let $\hat{\theta}_1$ and $\hat{\theta}_2$ be two CAN estimates of θ with asymptotic variances τ_1^2 and τ_2^2, respectively. The ratio τ_2^2/τ_1^2 is called the *asymptotic relative efficiency* (ARE) of $\hat{\theta}_1$ with respect to $\hat{\theta}_2$.

Least Squares. One advantage of least-squares regression is that the calculations are relatively simple and straightforward. Also, the least-squares estimates of the regression coefficients have several optimal properties. In the first place, by definition, they are the estimates that give the smallest sum of squared residuals.

Another optimal property is that the least-squares estimate of β_j is the best linear unbiased estimate (BLUE). That is, suppose we want to estimate β_j by a linear combination $a_1 y_1 + \cdots + a_n y_n$ of the observed response variables and suppose we want the estimate to be unbiased. Among all linear unbiased estimates of β_j, the one with the smallest variance is the least-squares estimate.

Now assume that the population of errors has a normal distribution. Then among all unbiased (not necessarily linear) estimates of β_j, the one with the smallest variance is the least-squares estimate. That is, the least-squares estimate of β_j is the uniformly minimum variance unbiased estimate (UMVUE). This property, however, depends strongly on the assumption of normality. When the distribution of the population of errors is not normal, the variance of the least-squares estimate can be much larger than the minimum possible variance.

In this and the next two paragraphs, we continue to assume that the population of errors has a normal distribution. An estimate $\hat{\beta}_j(\boldsymbol{y})$ is called equivariant if $\hat{\beta}_j(a\boldsymbol{y} + \boldsymbol{Xb}) = a\hat{\beta}_j(\boldsymbol{y}) + b_j$ for all a and $\boldsymbol{b}$. Among all equivariant estimates of β_j, the one with the smallest mean squared error, $E(\hat{\beta}_j - \beta_j)^2$, is the least-squares estimate.

The probability density function of $\boldsymbol{y}$ is $f(\boldsymbol{y}|\boldsymbol{\beta}, \sigma) = (2\pi\sigma^2)^{-n/2} \times \exp[-\|\boldsymbol{y} - \boldsymbol{X\beta}\|^2/(2\sigma^2)]$. Regarded as a function of the parameters $\boldsymbol{\beta}$ and σ for a fixed data vector $\boldsymbol{y}$, this is the likelihood function. The values of $\boldsymbol{\beta}$ and σ that maximize the likelihood are $\boldsymbol{\beta} = \hat{\boldsymbol{\beta}}_{\text{LS}}$ and $\sigma = (1/\sqrt{n})\|\boldsymbol{y} - \boldsymbol{X}\hat{\boldsymbol{\beta}}_{\text{LS}}\|$. That is, the least-squares estimate of β_j is the maximum likelihood estimate (MLE).

The F test of Section 3.8 is the uniformly most powerful invariant (UMPI) test. That is, among all tests of a fixed level that are invariant under a certain group of transformations of the data, the F test has the maximum power. When $q = p - 1$, it is also the uniformly most powerful unbiased (UMPU) test.

Suppose that the population of errors is not normally distributed. If we only assume that this population has a mean of 0, then the least-squares estimates may not be optimal but they at least have the property of being unbiased. If we further assume that the population of errors has finite variance, then the least-squares estimates have the property of being consistent and asymptotically normal, provided that the entries of the matrix $\boldsymbol{X}(\boldsymbol{X}'\boldsymbol{X})^{-1}\boldsymbol{X}'$ are all small.

A disadvantage of least-squares regression is that its optimality depends on willingness either to restrict attention to linear estimates or to assume that the random errors are normally distributed. When the distribution of the errors is nonnormal, least-squares estimates and tests may lose much of their efficiency. A few distant outliers can cause least-squares procedures to perform quite poorly.

Least Absolute Deviations. One optimal property of the LAD estimates of the regression coefficients is, by their definition, that they are the estimates that give the smallest sum of absolute residuals. Also, if we assume that the population of errors has a Laplace (or double exponential) distribution, then the LAD estimate of β_j is the maximum likelihood estimate.

If we assume only that there is a positive probability of the random errors being near 0, then the LAD estimates are consistent and asymptotically normal, provided that the entries of the matrix $(\boldsymbol{X}'\boldsymbol{X})^{-1}$ are all small. An indication of the stability of LAD estimation is given by the fact that their consistency does not require that the population of errors have finite variance.

The strength of LAD estimation is its robustness with respect to the distribution of the response variable (although not with respect to the explanatory variables). For this reason, LAD estimates are sometimes recommended as starting values for iterative estimation algorithms such as those in Chapters 5 and 6. The LAD method is especially suitable when it is thought that the distribution of the errors has very heavy tails or is asymmetric or when the sample is very large. (A distribution is said to have *heavy tails* if the graph of its probability density function appears to have thicker tails than the p.d.f. of a normal distribution. In other words, a heavy-tailed distribution produces a larger proportion of outliers than a normal distribution.) In the case of a heavy-tailed error distribution, LAD estimates have relatively low variance; in the case of an asymmetric error distribution, bias may be a more serious issue than variance and LAD estimates have low bias; in the case of a very large sample, any reasonable estimate has low variance and so it becomes more important to control bias.

For large samples, the efficiency of LAD estimates and tests relative to their LS counterparts is about the same as the efficiency of the sample median relative to the sample mean. If the errors have a normal distribution, the asymptotic relative efficiency (ARE) is only 0.64, but if the errors have a Laplace distribution, the ARE is 2. For the Cauchy distribution, the ARE is ∞. As long as the distribution of the error population is unimodal and symmetric about 0, the asymptotic relative efficiency of LAD with respect to least squares is at least 0.33. An ARE of 0.33 occurs for the uniform distribution, which is very short-tailed.

M-Regression. Consistency and asymptotic normality of Huber M-estimates have been proved under conditions requiring that all the entries in $\boldsymbol{X}(\boldsymbol{X}'\boldsymbol{X})^{-1}\boldsymbol{X}'$ are small, that n is very much larger than p, and that the error population is such that the Huber M-estimate is unbiased (which is true of any symmetric error population having a mean of 0).

The Huber M-estimate has minimax asymptotic variance among all M-estimates. To describe this property, recall that for any function $\rho(e)$, the corresponding M-estimate of $\boldsymbol{\beta}$ is defined to be the estimate that minimizes $\Sigma\rho(y_i - \boldsymbol{x}_i'\hat{\boldsymbol{\beta}})$. We only consider smooth, that is, differentiable, functions $\rho(e)$. Note that the least-squares estimate is the M-estimate corresponding to $\rho(e) = e^2$. The asymptotic variance of an estimate can be thought of as its variance for very large samples, multiplied by the sample size n. We would like our estimate to be robust in the sense that its asymptotic variance is relatively small not only if the population of errors has a normal distribution but also if it has a distribution that is close to normal. A large class of distributions that are "close" to normal is the class of ε-contaminated normal distributions, that is, distributions obtained by sampling a proportion $1 - \varepsilon$ of observations from a normal distribution and a proportion ε of observations from a contaminating distribution. We require the contaminating distribution to satisfy the condition that it does not induce any bias in the M-estimate. Suppose that for each M-estimate, we calculate its asymptotic variance for all ε-contaminated distributions and calculate the maximum of all these variances. The M-estimate with the smallest maximum asymptotic variance is the one corresponding to the Huber function $\rho(e)$ in (5.1). The value of ε determines the constant k. ($k = 1.5$ corresponds to $\varepsilon = 0.038$.)

For large samples, the efficiency of Huber M-estimates and tests relative to their LS counterparts is about the same as the efficiency of the Huber M-estimate relative to the sample mean in the case of a simple random sample (that is, with no explanatory variables). Efficiencies have been calculated for when σ is estimated using Huber's Proposal 2 (see Note 5.3g). If the errors have a normal distribution, the asymptotic relative efficiency (ARE) is 0.96. The ARE is high for error distributions that have heavy tails; for the Laplace distribution it is 1.31 and for the Cauchy distribution it is ∞. As long as the distribution of the error population is unimodal and symmetric about 0, the asymptotic relative efficiency of Huber M-regression with respect to least squares is at least 0.81.

Nonparametric Regression. The nonparametric rank-based estimates in Chapter 6 are consistent and asymptotically normal provided that all the entries of the matrix $(X'X)^{-1}$ are small and that the error population has a median of 0 and has finite Fisher information about location. (The normal, Laplace, and Cauchy distributions all have finite Fisher information about location.)

The asymptotic relative efficiency (ARE) of the rank-based estimates and tests in Chapter 6 with respect to their LS counterparts is the same as the ARE of the Wilcoxon signed-rank test with respect to the usual t test in the case of a simple random sample. If the errors have a normal distribution the ARE is 0.95, if the errors have a Laplace distribution it is 1.50, and if the errors have a Cauchy distribution it is ∞. As long as the distribution of the error population is symmetric about 0, the asymptotic relative efficiency with respect to least squares is at least 0.86.

Bayesian Regression. Those who advocate the use of Bayesian estimates and tests argue that the Bayesian approach to statistical analysis makes more sense than other approaches. When prior information is available, it seems wasteful not to use it, and a natural way to explicitly incorporate prior information into a statistical analysis is the Bayesian way. The Bayesian approach can also be regarded as being more satisfactory than the classical approach in that it produces a direct probability statement about a parameter or hypothesis, as opposed to the somewhat awkward notions of confidence level or p-value, which are frequently misinterpreted by nonprofessional statistical users. It could also be viewed as an advantage that Bayesian analysis allows one to interpret a probability as a measure of degree of belief concerning the actual observed data rather than as a long-run frequency involving hypothetical observations that might have been obtained but were not. Moreover, the Bayesian approach has the appealing feature that it provides a unified, fairly straightforward (conceptually if not computationally) way to analyze any statistical problem.

Critics of the Bayesian method say that it is too subjective, especially with regard to the choice of a prior distribution for the parameters. In response to this criticism, it can be said that statistical analysis cannot avoid being subjective, for example, in the choice of the model, and that the data analyst should admit his or her subjectivity and explicitly include it in the analysis. In doing this, however, one runs into the difficulty of quantifying one's prior knowledge and beliefs in the form of a prior distribution. Another difficulty arises if the prior distribution is not restricted to have a mathematically convenient form, because then the computation of the posterior distribution can be unwieldy.

As mentioned in Section 7.7, Bayesian regression estimates are the same as the least-squares estimates if a certain noninformative prior is used. For large sample sizes, most Bayesian procedures are approximately the same as the classical large-sample procedures.

Ridge Regression. When there are approximate linear relationships among the explanatory variables, ridge regression estimates are generally better than least-squares estimates, which may have large variances, be much too large, or have the wrong sign. In ridge regression the model is usually expressed as $\boldsymbol{y} = \mathbf{1}\mu + \boldsymbol{Z}\boldsymbol{\gamma} + \boldsymbol{e}$, where $\boldsymbol{Z}$ is the matrix of standardized explanatory variables. In the following paragraphs we consider ridge estimates of $\boldsymbol{\gamma}$ of the form (8.4). Unless stated otherwise, k is taken to be an arbitrary positive nonrandom number. In practice, k is a random quantity calculated from the data, but it is very difficult to derive properties of the ridge estimate when k is random.

First we look at the problem of least-squares estimates that are too large. Suppose we know that the length of $\boldsymbol{\gamma}$ should be no larger than an upper bound B. Recall that the least-squares estimate minimizes the length of the residual vector; that is, $\hat{\boldsymbol{\gamma}}_{\text{LS}}$ minimizes $\|\boldsymbol{y} - (\mathbf{1}\bar{y} + \boldsymbol{Z}\hat{\boldsymbol{\gamma}})\|$. If the length of $\hat{\boldsymbol{\gamma}}_{\text{LS}}$ is larger than B, then the ridge estimate, with k chosen so that its length $\|\hat{\boldsymbol{\gamma}}\|$ is equal to B, has the optimal property that it minimizes the length of the residual vector among all estimates $\hat{\boldsymbol{\gamma}}$ with length less than or equal to B.

Now consider the related problem of least-squares estimates that have large variance. The accuracy with which an estimate $\hat{\boldsymbol{\gamma}}$ estimates $\boldsymbol{\gamma}$ can be measured by its total mean squared error $E(\|\hat{\boldsymbol{\gamma}} - \boldsymbol{\gamma}\|^2)$. For an unbiased estimate like $\hat{\boldsymbol{\gamma}}_{\text{LS}}$, its total mean squared error is the sum of the variances of its components. It can be shown that there always exists an interval of small positive values of k for which the ridge estimate has smaller total mean squared error than the least squares estimate. (But unfortunately the interval depends on the values of the unknown parameters.) Under certain conditions on the eigenvalues of the matrix $\mathbf{Z}'\mathbf{Z}$, it has been found that the ridge estimate using the random value $k = p\hat{\sigma}^2_{\text{LS}}/\|\hat{\boldsymbol{\gamma}}_{\text{LS}}\|^2$, recommended in Section 8.6, has smaller total mean squared error than the least-squares estimate.

A ridge estimate can be viewed as a Bayesian estimate if we suppose $\boldsymbol{\gamma}$ is a random vector with mean vector $\mathbf{0}$ and covariance matrix $(\sigma^2/k)\boldsymbol{I}$. If the conditional distribution of $\boldsymbol{y}$ given $\boldsymbol{\gamma}$ and the distribution of $\boldsymbol{\gamma}$ are both assumed to be multivariate normal, then the posterior mean of $\boldsymbol{\gamma}$ is the ridge estimate. In an empirical Bayes approach, we can use the data to estimate the variances σ^2 and σ^2/k by $\hat{\sigma}^2_{\text{LS}}$ and $\|\hat{\boldsymbol{\gamma}}_{\text{LS}}\|^2/p$, respectively. Substituting these estimates into $k = \sigma^2/(\sigma^2/k)$, we obtain the ridge estimate of Section 8.6.

A ridge estimate can be viewed as the best linear unbiased estimate of the vector $\boldsymbol{\gamma}$ of random effects in the random linear model $\boldsymbol{y} = \mathbf{1}\mu + \boldsymbol{Z}\boldsymbol{\gamma} + \boldsymbol{e}$ in which $\boldsymbol{\gamma}$ has mean vector $\mathbf{0}$ and covariance matrix $(\sigma^2/k)\boldsymbol{I}$.

Caution. It must be kept in mind that the properties listed above depend on certain assumptions. In particular, the properties assume that the linear regression model is true and many of them assume that the error population has a particular kind of distribution. But our ultimate goal is to extract information from real data, and we cannot expect real data to exactly satisfy our assumptions. Also, many of the properties are asymptotic results; that is, they assume an infinitely large sample size, but of course all real samples are finite.

Summary. The principal features of these six types of regression can be summarized briefly as follows. Least-squares regression is optimal if the distribution of the error population is exactly normal but it can perform poorly if the distribution is not normal. Least-absolute-deviations regression is efficient when the error distribution has heavy tails and is effective at controlling bias. Huber M-regression and nonparametric rank-based regression are both efficient when the error distribution has heavy tails and do almost as well as least-squares when the error distribution is normal. Hence LAD, M-, and nonparametric regression are especially appropriate when the data contain outliers, since outliers are an indication of a heavy-tailed error distribution. Bayesian regression can be used to incorporate prior information into a statistical analysis. It produces direct probability statements about the parameters. Ridge regression may give more accurate estimates than least squares when there are collinearities among the explanatory variables.

9.3 COMPARISONS ON THREE DATA SETS

To get a more concrete idea of how the regression methods perform on real data, let us look at a few examples.

Example 1. Consider the acid content data in Table 3.1. In Figure 3.2 we see that all the data points fall closely around a straight line. For such a well-behaved data set, all the regression methods give very similar results. Estimates of α, β, and σ and the p-value for testing $\beta = 0$ were calculated using the methods presented in the preceding chapters. Compared to the least-squares estimates of α and β, the LAD estimates were within 2%, the M-estimates were exactly the same, the nonparametric estimates were within 1%, and the ridge estimates differed only in the fourth significant digit. Recall that, regardless of the data set, Bayesian estimates using the noninformative prior (7.2) are exactly the same as the least-squares estimates. The estimates of σ were not as close; they were 1.230, 1.433, 1.595, and 1.364 for least-squares, LAD, M-, and nonparametric regression, respectively. The p-values for testing $\beta = 0$ were all 0 to at least four decimal places and the Bayesian posterior probability of the null hypothesis was also approximately 0.

Example 2. Let us apply all six regression methods to the turnip green data in Section 1.2 using the linear regression model (3.9). Table 9.1 lists the estimates $\hat{\beta}_0, \hat{\beta}_1, \hat{\beta}_2, \hat{\beta}_3, \hat{\beta}_4$ of the regression coefficients, the estimate $\hat{\sigma}$ of the standard deviation of the error population, the number N_o of standardized residuals with absolute value larger than 2.5, the p-value P_{1234} for testing $\beta_1 = \beta_2 = \beta_3 = \beta_4 = 0$, the p-value P_1 for testing $\beta_1 = 0$, and the

Table 9.1 Results of Applying Various Regression Methods to the Turnip Green Data

	$\hat{\beta}_0$	$\hat{\beta}_1$	$\hat{\beta}_2$	$\hat{\beta}_3$	$\hat{\beta}_4$
Least squares	119.6	−0.03367	5.425	−0.5026	−0.1209
LAD	133.8	−0.05301	6.635	−0.6974	−0.1460
M-regression	122.7	−0.03967	5.763	−0.5443	−0.1282
Nonparametric	123.7	−0.04478	6.043	−0.5583	−0.1339
Ridge	115.9	−0.02805	4.807	−0.4363	−0.1089
	$\hat{\sigma}$	N_o	P_{1234}	P_1	P_3
Least squares	6.104	0	0.0000	0.1811	0.0263
LAD	4.140	4	0.0000	0.0149	0.0012
M-regression	4.177	4	0.0000	0.0426	0.0030
Nonparametric	4.509	3	0.0000	0.0341	0.0027

p-value P_3 for testing $\beta_3 = 0$. For the ridge regression method, only the estimates of the regression coefficients are given because this method is intended to be used only for estimation. The Bayesian method is not included in the table because, using the noninformative prior distribution (7.7), the Bayesian estimates are the same as the least-squares estimates, and the Bayesian method does not define p-values. Using the noninformative prior distribution (7.14), the Bayesian posterior probabilities of the three hypotheses are 0.0000, 0.6781, and 0.2956.

The five estimation methods in the table produce estimated coefficients that are noticeably different although "in the same ballpark".

According to the discussion in the preceding section, LAD, M-, and nonparametric regression are especially suitable when there are outliers in the data. In all three procedures the data points numbered 10, 19, and 20 had standardized residuals more than 2.5 in absolute value and hence may be regarded as outliers. Point number 15 was also detected as an outlier by LAD and M-regression and was almost detected by the nonparametric procedure. These points should not necessarily be omitted from the data set, but to see their effect on the analysis we applied least-squares, LAD, M-, and nonparametric regression with the four outliers omitted. Without the outliers, the estimates from the four procedures were closer to one another. The stability of the LAD procedure was seen in the fact that the LAD estimates were almost unchanged when the outliers were omitted.

Ridge regression is especially suitable when there is collinearity among the explanatory variables. Because of the high correlation of 0.997 between X_2 and X_4 $(= X_2^2)$, we expect ridge regression to give more accurate estimates of β_2 and β_4 than least squares. Note that the ridge estimates $\hat{\beta}_1$, $\hat{\beta}_2$, $\hat{\beta}_3$, and $\hat{\beta}_4$ are all closer to 0 than the corresponding least-squares estimates. This agrees with the description in Chapter 8 of ridge regression as a procedure that shrinks the least-squares estimates.

When the evidence against a hypothesis is very strong, as it is for the hypothesis that $\beta_1 = \beta_2 = \beta_3 = \beta_4 = 0$, then all four non-Bayesian testing procedures yield a p-value near 0 and the Bayesian procedure yields a posterior probability near 0. For the other two hypotheses the p-value for least-squares is much larger than the p-values for the other three procedures. Outliers can have a strong effect on least-squares tests. When the four outliers were omitted, the p-value for least-squares became closer to the p-values for the other procedures.

Example 3. Next we consider a data set that has appeared as an example in many books and articles. The data, displayed in Table 9.2, consist of measurements from a factory for the oxidation of ammonia to nitric acid. On 21 different days, measurements were taken of the air flow (X_1), the

Table 9.2 Stack Loss Data

Stack Loss (Y)	Rate (X_1)	Temperature (X_2)	Acid Concentration (X_3)
42	80	27	89
37	80	27	88
37	75	25	90
28	62	24	87
18	62	22	87
18	62	23	87
19	62	24	93
20	62	24	93
15	58	23	87
14	58	18	80
14	58	18	89
13	58	17	88
11	58	18	82
12	58	19	93
8	50	18	89
7	50	18	86
8	50	19	72
8	50	19	79
9	50	20	80
15	56	20	82
15	70	20	91

Source: Brownlee (1965, p. 454).

temperature of cooling water (X_2), the concentration of acid (X_3), and the amount of ammonia that escaped before being oxidized, called stack loss (Y).

All six regression methods were applied using the model $Y = \beta_0 + \beta_1 X_1 + \beta_2 X_2 + \beta_3 X_3 + e$. Table 9.3 shows the estimates $\hat{\beta}_0$, $\hat{\beta}_1$, $\hat{\beta}_2$, $\hat{\beta}_3$, and $\hat{\sigma}$, the number N_o of standardized residuals with absolute value larger than 2.5, and the p-values P_1, P_2, P_3, and P_{123} for testing the hypotheses $\beta_1 = 0$, $\beta_2 = 0$, $\beta_3 = 0$, and $\beta_1 = \beta_2 = \beta_3 = 0$. The Bayesian posterior probabilities of the four hypotheses, using the noninformative prior distribution (7.14), are 0.0014, 0.0436, 0.7465, and 0.0000.

There are substantial differences in the estimates of β_1, β_2, and β_3 for the five methods. This is at least partly due to outliers. As in Example 2, the M- and nonparametric estimates are similar to one another.

In an extensive least-squares analysis of this data set, Daniel and Wood (1980, Chapter 5) found four outliers, the observations numbered 1, 3, 4, and 21. In LAD regression, three of these outliers had standardized residuals larger than 2.5 in absolute value and the other outlier had a standardized residual of 2.331. M- and nonparametric regression, however, detect only one outlier and almost detect another one. When we applied the five estimation

Table 9.3 Results of Applying Various Regression Methods to the Stack Loss Data

	$\hat{\beta}_0$	$\hat{\beta}_1$	$\hat{\beta}_2$	$\hat{\beta}_3$	$\hat{\sigma}$
Least squares	−39.92	0.7156	1.295	−0.1521	3.243
LAD	−39.69	0.8319	0.574	−0.0609	2.171
M-regression	−41.17	0.8133	1.000	−0.1324	2.661
Nonparametric	−40.16	0.8155	0.888	−0.1202	2.920
Ridge	−40.62	0.6861	1.312	−0.1273	
	N_o	P_1	P_2	P_3	P_{123}
Least squares	0	0.0001	0.0026	0.3440	0.0000
LAD	3	0.0001	0.0211	0.2921	0.0000
M-regression	1	0.0000	0.0114	0.3183	0.0000
Nonparametric	1	0.0001	0.0130	0.4000	0.0000

methods to the data with the four outliers omitted, the five estimates of each coefficient were more similar to one another than they are in Table 9.3, all being roughly similar to the LAD estimate from the entire data set.

The p-values for the four testing methods in Table 9.3 lead to the same conclusions, except possibly for P_2. The methods still differed in their conclusions about the hypothesis $\beta_2 = 0$ when the outliers were omitted.

Recommendations. In each example above we have applied six different methods of regression to the same set of data for the purpose of comparing the methods. If our only purpose had been to analyze the data, it would still be good practice to apply more than one (but maybe not as many as six) regression methods. If you use several methods to analyze a data set and they all lead to similar results, you can feel confident about your conclusions. If there are serious disagreements between the results of the different methods, you should examine the data to see why.

We would recommend using least squares and one other method. The personal preferences of the authors are the nonparametric rank-based and LAD methods, but any of the methods presented in this book are justifiable. Each method has its advantages and disadvantages. More important than choosing the method is applying it responsibly, that is, not automatically but interactively, with attention to the particular features of the data at hand.

NOTES

9.2a. Asymptotic properties are expressed in terms of the limiting behavior of a sequence of estimates calculated from samples of increasing size, rather than the behavior of a single estimate calculated from a sample of a given size. In practice, we are interested in a single estimate. Asymptotic properties are relevant for this estimate to the extent that the given sample size is "large" and the sequence of estimates is a "natural" extension of the estimate of interest.

9.2b. More precise definitions of consistency and asymptotic normality may be found, for example, in Lehmann (1983, pp. 332, 340).

9.2c. The asymptotic relative efficiency of one estimate with respect to another estimate can be defined in terms of sample sizes. Let $\hat{\theta}_1$ and $\hat{\theta}_2$ be two estimates of a parameter θ. To designate the size of the sample from which the estimates are calculated, we can use the notation $\hat{\theta}_{1,n}$ and $\hat{\theta}_{2,n}$. For each n, let n' be the sample size such that the variance of $\hat{\theta}_{2,n'}$ is equal (or as close as possible) to the variance of $\hat{\theta}_{1,n}$. The *asymptotic relative efficiency of* $\hat{\theta}_1$ with respect to $\hat{\theta}_2$ is the limit of n'/n as n becomes infinitely large. It can also be expressed as the reciprocal of the ratio of their asymptotic variances. See Lehmann (1983, Section 5.2).

9.2d. The asymptotic relative efficiency of one test with respect to another test is defined similarly to Note 9.2c. Let T_1 and T_2 be two tests of a hypothesis H. For each n, for a given significance level, let n' be the sample size such that the power of $T_{2,n'}$ is equal (or as close as possible) to the power of $T_{1,n}$ against an alternative close to H. The *asymptotic relative efficiency* of T_1 with respect to T_2 is the limit of n'/n as n becomes infinitely large. Under certain conditions, the limit does not depend on the particular level or power that is considered. See Lehmann (1975, Appendix, Section 6).

9.2e. Properties of the least-squares estimates, namely, that they are BLU and, under the assumption of normality, are UMVU, best equivariant (sometimes called best invariant), and ML, are derived in Arnold (1981, Chapter 6). See Note 3.7 for a verification of unbiasedness. Consistency and asymptotic normality are shown in Huber (1981, Section 7.2). The fact that the F test is UMPI under normality is derived in Arnold (1981, Section 7.6) and Lehmann (1986, Section 7.1). For the UMPU property when $q = p - 1$, see Lehmann (1986, Section 7.1 and Problem 5 in Chapter 5).

9.2f. The loss of efficiency of least-squares estimates for mildly nonnormal error distributions is discussed in Hampel et al. (1986, Section 1.2d).

9.2g. Properties of LAD estimates are discussed in Bloomfield and Steiger (1983, Preface and Section 2.2) and Huber (1987). Bloomfield and Steiger show the consistency and asymptotic normality of LAD estimates when the explanatory variables are regarded as random (see their Theorems 1 and 2 in Section 2.2). In this book, however, we regard the explanatory variables as nonrandom. Consistency and asymptotic normality in this case are proved by

Bassett and Koenker (1978) (note that if $\hat{\boldsymbol{\beta}}$ is asymptotically normal with asymptotic mean $\boldsymbol{\beta}$, then this implies that $\hat{\boldsymbol{\beta}}$ is consistent with $\boldsymbol{\beta}$).

9.2h. LAD estimates seem to have lower bias than other estimates of the regression function when the population of errors is asymmetrically distributed. Since we assume that the population of errors has a mean of 0, the regression function is a population mean. When a population is symmetrically distributed around some point, then the mean and median, as well as other candidates for the "central" point of the distribution, coincide with the point of symmetry. But when a population is asymmetrically distributed, then these various "central" points differ and it is not clear which one we should estimate.

9.2i. Asymptotic relative efficiencies of the median with respect to the mean for various distributions of the error population may be found in Lehmann (1983, Section 5.3).

9.2j. A distribution is *unimodal* at 0 if it has a probability density function $f(x)$ that is nondecreasing for x less than 0 and is nonincreasing for x greater than 0.

9.2k. Asymptotic properties of M-regression estimates are derived in Huber (1981, Section 7.4). The minimax property of the Huber M-estimate is derived in Chapter 4 of the same book for a simple random sample with no explanatory variables, and according to Huber (1981, p. 74) the result carries over to regression. See Hampel et al. (1986, Section 8.2a) for a discussion of Huber's minimax approach to estimation. The asymptotic behavior of M-regression tests is considered in Schrader and Hettmansperger (1980).

9.2l. Asymptotic relative efficiencies of the Huber M-estimate with respect to the mean for various distributions of the error population may be found in Bickel (1965).

9.2m. A Huber M-estimate would be a maximum likelihood estimate if the error distribution had a probability density function of the form $f(u) = Ce^{-\rho(u)}$.

9.2n. Asymptotic properties of nonparametric rank-based estimates and tests are derived in Hettmansperger (1984, Chapter 5).

9.2o. Asymptotic relative efficiencies of the Wilcoxon signed-rank test with respect to the least-squares t test for various distributions of the error population may be found in Hettmansperger (1984, Section 2.6) and Lehmann (1983, Section 5.6).

9.2p. Arguments for and against the Bayesian approach are given in Berger (1985, Sections 3.7 and 4.1), Press (1989, Section 2.12), and Barnett (1982, Section 6.8.3).

9.2q. The property mentioned in the second paragraph of the subsection on ridge regression is given in Hoerl and Kennard (1970, Sections 3b and 6). The property for nonrandom k in the third paragraph and the Bayesian view of ridge estimation are given in Theorem 4.3 and in Section 6 of the same article. The property for random k mentioned in the third paragraph may be found in Alam and Hawkes (1978). For best linear unbiased estimation (or prediction) of random effects, see Henderson (1975, Section 2).

9.2r. Most of the research on LAD, M-, and nonparametric regression has dealt with performance for large samples and for symmetric error distributions. Fewer results are available about their performance for small samples or for asymmetric error distributions, but see Schrader and McKean (1987), Field and Ronchetti (1991), and McKean and Sheather (1991).

9.3a. The acid content data are "well behaved" in the sense that they are quite compatible with the assumptions of the simple linear regression model. However, among regression data sets with more than one explanatory variable, such good behavior is much rarer. See Daniel and Wood (1980, p. 60).

9.3b. In Table 9.1, the p-values for the LAD procedure were obtained using the F distribution rather than the chi-squared distribution. The p-values using the chi-squared distribution were smaller.

9.3c. Estimation methods have been developed for dealing with regression data that contain both collinearity and outliers. See Pfaffenberger and Dielman (1990).

9.3d. Three of the four outliers detected by LAD and M-regression in the turnip green data can also be detected using least-squares diagnostic procedures. This requires a more sophisticated procedure than simply looking at the LS residuals divided by $\hat{\sigma}_{LS}$. We have mentioned studentized residuals in Note 2.4a. Also, in order to judge whether a data point is an outlier, it is a good precaution to eliminate its influence on the estimates $\hat{\boldsymbol{\beta}}$, and $\hat{\sigma}$ that are used in forming its studentized residual. This is the idea behind the *externally studentized residual* (see Weisberg, 1985, Section 5.2). The externally studentized residual of a data point can be used to test the null hypothesis that the point is not an outlier. The p-values for data points 10, 15, 19, and 20 are 0.0343, 0.0329, 0.0111, and 0.1157.

9.3e. The collinearity problem in the turnip green data can be reduced by standardizing X_2 before forming its square. We can replace X_2 and X_4 by $Z_2 = (X_2 - 18.8)/20.71$ and $Z_4 = Z_2^2$, where 18.8 and 20.71 are the mean and standard deviation of the 27 observed values of X_2. Since Z_2 and Z_4 are linear functions of X_2 and X_4, the linear regression model with explanatory variables X_1, Z_2, X_3, and Z_4 is essentially equivalent to model (3.9). In the modified model the correlation between Z_2 and Z_4 is 0.954, which is less than 0.997.

9.3f. Table 9.3 is similar to Table 3.1 in Dodge (1984). The LS and LAD residuals are displayed in the earlier table. The Huber estimates in the two tables are different because the earlier table used $k = 1.345$ rather than $k = 1.5$.

9.3g. Analysis of a data set by more than one method is recommended by Hogg (1979) and Tukey (1991).

REFERENCES

Alam, K., and J. S. Hawkes (1978). Estimation of regression coefficients. *Scandanavian Journal of Statistics*, vol. 5, pp. 169–172.

Arnold, S. F. (1981). *The Theory of Linear Models and Multivariate Analysis*. Wiley, New York.

Barnett, V. (1982). *Comparative Statistical Inference*, 2nd ed. Wiley, New York.

Bassett, G. Jr., and R. Koenker (1978). Asymptotic theory of least absolute error regression. *Journal of the American Statistical Association*, vol. 73, pp. 618–622.

Berger, J. O. (1985). *Statistical Decision Theory and Bayesian Analysis*, 2nd ed. Springer-Verlag, New York.

Bickel, P. J. (1965). On some robust estimates of location. *Annals of Mathematical Statistics*, vol. 36, pp. 847–858.

Bloomfield, P., and W. L. Steiger (1983). *Least-Absolute Deviations: Theory, Applications, and Algorithms*. Birkhäuser, Boston.

Brownlee, K. A. (1965). *Statistical Theory and Methodology in Science and Engineering*, 2nd ed. Wiley, New York.

Daniel, C., and F. S. Wood (1980). *Fitting Equations to Data*, 2nd ed. Wiley, New York.

Dodge, Y. (1984). Robust estimation of regression coefficients by minimizing a convex combination of least squares and least absolute deviations. *Computational Statistics Quarterly*, vol. 1, pp. 139–153.

Field, C. A., and E. M. Ronchetti (1991). An overview of small sample asymptotics. In: W. Stahel and S. Weisberg (eds.), *Directions in Robust Statistics and Diagnostics, Part I*. Springer-Verlag, New York.

Hampel, F. R., E. M. Ronchetti, P. J. Rousseeuw, and W. A. Stahel (1986). *Robust Statistics: The Approach Based on Influence Functions*. Wiley, New York.

Henderson, C. R. (1975). Best linear unbiased estimation and prediction under a selection model. *Biometrics*, vol. 31, pp. 423–447.

Hettmansperger, T. P. (1984). *Statistical Inference Based on Ranks*. Wiley, New York.

Hoerl, A. E., and R. W. Kennard (1970). Ridge regression: biased estimation for nonorthogonal problems. *Technometrics*, vol. 12, pp. 55–67.

Hogg, R. V. (1979). Statistical robustness: one view of its use in applications today. *American Statistician*, vol. 33, pp. 108–115.

Huber, P. J. (1981). *Robust Statistics*. Wiley, New York.

Huber, P. J. (1987). The place of the L_1-norm in robust estimation. In: Y. Dodge (ed.), *Statistical Data Analysis Based on the L_1-Norm and Related Methods*. North-Holland, New York.

Lehmann, E. L. (1975). *Nonparametrics: Statistical Methods Based on Ranks*. Holden-Day, San Francisco.

Lehmann, E. L. (1983). *Theory of Point Estimation*. Wiley, New York.

Lehmann, E. L. (1986). *Testing Statistical Hypotheses*, 2nd ed. Wiley, New York.

McKean, J. W., and S. J. Sheather (1991). Small sample properties of robust analyses of linear models based on R-estimates: a survey. In: W. Stahel and S. Weisberg (eds.), *Directions in Robust Statistics and Diagnostics, Part II*. Springer-Verlag, New York.

Pfaffenberger, R. C., and T. E. Dielman (1990). A comparison of regression estimators when both multicollinearity and outliers are present. In: K. D. Lawrence and J. L. Arthur (eds.), *Robust Regression: Analysis and Applications*. Marcel Dekker, New York.

Press, S. J. (1989). *Bayesian Statistics: Principles, Models, and Applications*. Wiley, New York.

Schrader, R. M., and J. W. McKean (1987). Small sample properties of least absolute errors analysis of variance. In: Y. Dodge (ed.), *Statistical Data Analysis Based on the L_1-Norm and Related Methods*. North-Holland, New York.

Schrader, R. M., and T. P. Hettmansperger (1980). Robust analysis of variance based on a likelihood criterion. *Biometrika*, vol. 67, pp. 93–101.

Tukey, J. W. (1991). Graphical displays for alternate regression fits. In: W. Stahel and S. Weisberg (eds.), *Directions in Robust Statistics and Diagnostics, Part II*. Springer-Verlag, New York.

Weisberg, S. (1985). *Applied Linear Regression*, 2nd ed. Wiley, New York.

CHAPTER 10

Other Methods

Alas, that Spring should vanish with the Rose
That Youth's sweet-scented Manuscript should close
The Nightingale that in the Branches sang,
Ah, whence, and whither flown again, who knows.

KHAYYAM NAISHAPURI

10.1 INTRODUCTION

There are many other linear regression methods besides the six presented in this book. Many variations of these six methods exist and some totally different methods have been invented. In this final chapter we mention a few of these. In the last section we briefly consider some generalizations of the linear regression model.

10.2 OTHER METHODS OF LINEAR REGRESSION

Compromises Between LS and LAD Estimation. In least-squares and least-absolute-deviations regression the estimates of the regression coefficients are chosen to minimize $\Sigma\hat{e}_i^2$ and $\Sigma|\hat{e}_i|$, respectively. More generally, the estimates can be chosen to minimize $\Sigma|\hat{e}_i|^p$, the sum of the pth powers of the absolute residuals. This is called *L_p-norm estimation*. A suitable value of p can be determined from the data. See Gonin and Money (1989). As p approaches infinity, this leads to L_∞-norm estimation, which minimizes the maximum absolute residual.

For a value of p between 1 and 2, L_p-norm estimation can be regarded as a compromise between least-squares and LAD estimation. It is a kind of M-estimation, minimizing $\Sigma\rho(\hat{e}_i)$, where the function $\rho(e)$ is a compromise between e^2 and $|e|$. Huber M-estimation, described in Chapter 5, uses a function $\rho(e)$ that is equal to e^2 for values of e close to 0 and is a linear function of $|e|$ for values of e far from 0. Another such compromise is M-estimation using a function $\rho(e)$ that is a weighted average of the functions e^2 and $|e|$; see Dodge (1984).

Other M-Estimates. An M-estimate is often presented in terms of the function $\psi(e) = \rho'(e)$ rather than the function $\rho(e)$. This is sufficient because the M-estimate can be found by solving $\Sigma\psi(y_i - \boldsymbol{b}'\boldsymbol{x}_i)\boldsymbol{x}_i = 0$ rather than minimizing $\Sigma\rho(y_i - \boldsymbol{b}'\boldsymbol{x}_i)$. (Recall that, in the case of simple regression, we can solve (5.3) rather than minimize (5.2). As in Chapter 5, the notation $\boldsymbol{x}_i$ denotes the vector $(1, x_{i1}, \ldots, x_{ip})$.) For Huber M-estimation, the function $\psi(e)$ is constant with value $-2k$ for $e \leq -k$, is a linear function $2e$ for $-k \leq e \leq k$, and is constant with value $2k$ for $k \leq e$. The ψ-function is especially interesting because, when multiplied by the appropriate constant, it coincides with the residual influence function of the estimate.

The *influence function* of an estimate measures the amount of bias in the estimate that would be caused by a small amount of contamination in the data. The influence function can be written as a product of the influence of the residual, which indicates the effect of contamination in the response variable (or in the errors), and the influence of position, which indicates the effect of contamination in the explanatory variables. See Hampel et al. (1986, Sections 4.2a and 6.2).

In order to be robust, an M-estimate should have a bounded ψ-function, so as to bound the influence of outlying values of the response variable. Note that the Huber ψ-function is bounded between $-2k$ and $2k$. The ψ-function for the least-squares estimate is $\psi(e) = 2e$ for all e, which is unbounded. The influence of outlying values of the response variable can be eliminated completely by choosing a ψ-function such that $\psi(e) = 0$ when e is far from 0. An estimate obtained in this way is called a *redescending M-estimate*. See Hampel et al. (1986, Section 2.6).

A bounded ψ-function, however, cannot prevent undue influence from leverage points, that is, outlying values of the explanatory variables. To solve this problem, generalized M-estimates were introduced. A *Mallows generalized M-estimate* is obtained by solving $\Sigma w(\boldsymbol{x}_i)\psi(y_i - \boldsymbol{b}'\boldsymbol{x}_i)\boldsymbol{x}_i = 0$, where $w(\boldsymbol{x})$ is a nonnegative function. By choosing a suitable weight function $w(\boldsymbol{x})$, one can bound the influence of the vector $\boldsymbol{x}$ of explanatory variables. Such an estimate is also called a *bounded-influence estimate*. See Hampel et al. (1986, Sections 6.3a and 6.3b).

L-Estimates. The three basic types of robust estimation that Huber included in his 1981 book are M-, L-, and R-estimation. In the case of a simple random sample without explanatory variables, an *L-estimate* is a linear combination of order statistics, such as a trimmed mean. Note that a trimmed mean can be viewed as another compromise between LS and LAD estimation. The LS estimate in a simple random sample is the mean, which is a "trimmed" mean with 0% trimming, and the LAD estimate is the median, which is a trimmed mean with 100% trimming. L-estimation has been extended to regression data by extending the notion of an order statistic to the notion of a regression quantile; see Koenker and Basset (1978) and Koenker and Portnoy (1987).

R-Estimates. The "R" in R-estimation stands for "rank-based". The nonparametric estimates presented in Chapter 6 are R-estimates. They are chosen to minimize (6.6), $\Sigma[\text{rank}(\hat{e}_i) - \frac{1}{2}(n+1)]\hat{e}_i$. More generally, *R-estimates* can be chosen to minimize $\Sigma a(r_i)\hat{e}_i$, where r_i denotes rank($\hat{e}_i$) and $a(r)$ is a function called a score function. In Chapter 6 we have used the Wilcoxon score function, $a(r) = r - \frac{1}{2}(n+1)$. Other score functions that are sometimes used are the normal score function, $a(r) = \Phi^{-1}(r/(n+1))$, where $\Phi(t)$ is the cumulative distribution function of the standard normal distribution, and the sign score function, $a(r) = \text{sgn}(r - \frac{1}{2}(n+1))$, where sgn($t$) is the sign of t (-1 if t is negative, 0 if $t = 0$, and 1 if t is positive). See Hettmansperger (1984, Section 5.2) and McKean and Sievers (1989).

High-Breakdown-Point Estimates. The robustness of an estimate against a single outlier is measured by its influence function. The robustness of an estimate against heavier contamination is measured by its *breakdown point*, which is the largest proportion of outliers that can occur in a sample without entailing the possibility of arbitrarily large bias. See Rousseeuw and Leroy (1987, Section 1.2). The LS, LAD, and Huber M-estimates are not robust in this regard; they have breakdown points of 0% because of their sensitivity to leverage points. The rank-based estimate in Chapter 6 has a breakdown point less than 30%.

The maximum possible breakdown point is 50%. This is achieved by the *least-median-of-squares* (LMS) estimate, which is the estimate that minimizes the median of the squared residuals $\hat{e}_i^2$ (or, equivalently, minimizes the median of the absolute residuals $|\hat{e}_i|$). See Rousseeuw and Leroy (1987). In the plot of a simple regression data set, the LMS line is the middle of the narrowest strip containing half of the data points. The LMS estimate is simple to describe and is very robust against outliers. For certain types of data sets, however, the estimate is not robust against small changes in centrally located data points. LMS estimation involves considerable computation and has low efficiency, but it can be useful in the detection of outliers or as a robust initial estimate for iterative procedures. More complicated regression estimates have been devised that have breakdown points of 50% and also have efficiencies as high as 0.95 for normally distributed errors, but computation is still a problem. See Jurečková and Portnoy (1987) and Yohai and Zamar (1988).

Shrinkage Estimates. The ridge estimate of $\boldsymbol{\gamma}$, the vector of regression coefficients of the standardized explanatory variables, is "shrunk" in the sense that its length is less than the least-squares estimate. Now suppose that the explanatory variables $X_1, \ldots, X_q$ are thought to be more important than $X_{q+1}, \ldots, X_p$. Rather than shrink all the regression coefficient estimates, one may prefer to shrink the estimates for only the unimportant variables. This is the idea of *subspace-ridge estimation*; see Oman (1982) and Lee and Birkes (1993).

Another type of shrinkage estimate is the Stein estimate. As with ridge estimation, it is common to standardize the explanatory variables before applying Stein estimation. The *Stein estimate* of $\boldsymbol{\gamma}$ is $c\hat{\boldsymbol{\gamma}}_{\mathrm{LS}}$, where $c = [1 - (a/F)]^+$, $F = \|\mathbf{Z}\hat{\boldsymbol{\gamma}}_{\mathrm{LS}}\|^2/p\hat{\sigma}^2_{\mathrm{LS}}$, $a = (1 - 2/p)(1 - 2/(n - p + 1))$, and $[u]^+ = \max(u, 0)$. See Rolph (1976). It is applicable only when $p \geq 3$. To see why this is a sensible estimate, note that F is the least-squares test statistic for testing the hypothesis $\boldsymbol{\gamma} = \mathbf{0}$. If the hypothesis is true, then F will probably be close to 1, which implies c will be close to 0 and hence the Stein estimates of $\boldsymbol{\gamma}$ will be close to $\mathbf{0}$. If the hypothesis is false, then F will probably be large, which implies c will be close to 1 and the Stein estimate will be close to the least-squares estimate.

If some of the explanatory variables are thought to be more important than others, one can use a *subspace-Stein estimate*, which shrinks the coefficient estimates of only the unimportant variables; see Jennrich and Oman (1986). Subspace-Stein estimation can be viewed as a "smooth" version of *preliminary-test estimation*, in which estimation is preceded by a test of $\beta_{q+1} = \cdots = \beta_p = 0$. If the hypothesis is rejected, then the ordinary least-squares estimates of the regression coefficients are used. If the hypothesis is accepted, then $\beta_{q+1}, \ldots, \beta_p$ are estimated to be 0 and $\beta_1, \ldots, \beta_q$ are estimated by least squares in the reduced model. Rather than choose between least squares in the full model and least squares in the reduced model on the basis of a test statistic F, subspace-Stein estimation takes a weighted average of the two, with the weights depending on the value of F.

Principal Components Regression. To deal with regression data having collinearities among the explanatory variables, another technique one can use, besides ridge regression, is principal components regression. The first *principal component* is the standardized linear combination of the explanatory variables that shows the most variability. (A linear combination is standardized if the sum of the squares of its coefficients is equal to 1.) Among all standardized linear combinations of the explanatory variables that are orthogonal to the first principal component, the second principal component is the one showing the most variability. Continuing in this way, p principal components can be constructed. Sometimes, by looking at the sizes and signs of the coefficients in the linear combination, a meaningful interpretation can be given to a principal component. Explanatory variables that are collinear with one another will typically occur together with large coefficients in a single principal component. In *principal components regression*, the first several principal components are taken as a new set of explanatory variables and least-squares estimates are calculated. See Gunst and Mason (1980, Section 10.1) and Myers (1990, Section 8.4).

Maximum Likelihood Estimation. If one is willing to assume that the error population has a particular kind of distribution, maximum likelihood (ML) estimation can be applied. The error distribution is usually assumed to be continuous, which means that it can be described in terms of a probability density function (p.d.f.). The exact error distribution is almost always unknown to us, but suppose we are willing to assume that the distribution is contained in a known family of distributions with p.d.f.'s $f(e;\sigma)$ indexed by the parameter σ. (More generally, the family could be indexed by a vector of parameters.) Then the p.d.f. of the sample of observed response variables is the product $f(y_1 - \boldsymbol{\beta}'\mathbf{x}_1;\sigma)\cdots f(y_n - \boldsymbol{\beta}'\mathbf{x}_n;\sigma)$. Regarded as a function of $\boldsymbol{\beta}$ and σ, this is the likelihood function. The *maximum likelihood estimates* of $\boldsymbol{\beta}$ and σ are chosen to maximize the likelihood function.

If the family of possible error distributions is the family of normal distributions with mean 0, then the ML estimates of the regression coefficients coincide with the least-squares estimates; the ML estimate of σ is $(\sqrt{(n-p-1)/n}\,)\hat{\sigma}_{\mathrm{LS}}$. For the family of Laplace (double exponential) distributions with mean 0, the ML estimates of the regression coefficients coincide with the LAD estimates.

10.3 MORE GENERAL METHODS OF REGRESSION

For some sets of regression data a linear regression model may not be appropriate. The techniques mentioned in Chapter 2 may fail to produce a satisfactory model. Or theory about the process that generated the data may suggest a different type of model. Such data require more general models.

A very general model is $y_i = \mu_i + e_i$, where μ_i is the expectation of the random variable y_i and where e_i is defined to be the difference (or "error")

$y_i - \mu_i$. Different kinds of models are obtained by making different assumptions about the structure of the expectations μ_i and about the distribution of the errors e_i. The linear regression model assumes that μ_i is a linear function $\boldsymbol{\beta}'\boldsymbol{x}_i$ of the explanatory variables and that the random variables e_i are independent of one another and have the same distribution.

Weighted Least Squares. For some regression data sets it may be valid to assume that $\mu_i = \boldsymbol{\beta}'\boldsymbol{x}_i$ but invalid to assume that the errors all come from the same population. Suppose that the errors are independent but have unequal standard deviations that are proportional to one of the explanatory variables, say, X_1, so that $\text{Var}(e_i) = \sigma^2 x_{i1}^2$. Or more generally, suppose $\text{Var}(e_i) = \sigma^2 v_i$ for known positive quantities v_i. The *weighted least-squares estimate* $\hat{\boldsymbol{\beta}}_{\text{WLS}}$ is the value of $\boldsymbol{\beta}$ that minimizes $\Sigma(y_i - \boldsymbol{\beta}'\boldsymbol{x}_i)^2/v_i$. This becomes the ordinary least-squares procedure in the special case when all the v_i's are equal. See Weisberg (1985, Chapter 4) and Myers (1990, Section 7.1).

Generalized Least Squares. Continue to assume that the expectation of y_i is a linear function of the explanatory variables. In vector notation, $\boldsymbol{y} = \boldsymbol{X}\boldsymbol{\beta} + \boldsymbol{e}$. In ordinary least-squares regression the vector of errors is assumed to have the variance–covariance matrix $\text{Cov}(\boldsymbol{e}) = \sigma^2\boldsymbol{I}$. In weighted least squares it is assumed that $\text{Cov}(\boldsymbol{e}) = \sigma^2\boldsymbol{V}$, where $\boldsymbol{V}$ is a diagonal matrix with positive diagonal entries. More generally, suppose $\text{Cov}(\boldsymbol{e}) = \sigma^2\boldsymbol{V}$, where $\boldsymbol{V}$ is any invertible variance–covariance matrix. This allows the errors to be correlated as well as to have unequal variances. The *generalized least-squares estimate* of the vector of regression coefficients is $\hat{\boldsymbol{\beta}}_{\text{GLS}} = (\boldsymbol{X}'\boldsymbol{V}^{-1}\boldsymbol{X})^{-1}\boldsymbol{X}'\boldsymbol{V}^{-1}\boldsymbol{y}$. When $\boldsymbol{V} = \boldsymbol{I}$, this reduces to formula (3.8). See Myers (1990, Section 7.1) and Seber (1977, Section 3.6).

Nonlinear Regression. A linear regression function is not always suitable. Consider a regression function of the general form $\mu_i = g(\boldsymbol{x}_i, \boldsymbol{\theta})$, where g is a known function that is not necessarily linear and $\boldsymbol{\theta}$ is a vector of unknown parameters. (In nonlinear regression no convenience is gained by including 1 as the first component of $\boldsymbol{x}_i$ and so we use the more natural notation $\boldsymbol{x}_i = (x_{i1}, \ldots, x_{ip})$.) The least-squares estimate $\hat{\boldsymbol{\theta}}_{\text{LS}}$ is the value of $\boldsymbol{\theta}$ that minimizes $\Sigma[y_i - g(\boldsymbol{x}_i, \boldsymbol{\theta})]^2$. When g is a nonlinear function, there is generally no explicit formula for $\hat{\boldsymbol{\theta}}_{\text{LS}}$; it must be computed iteratively. One possible iterative algorithm is the following. First pick an initial estimate. Using a Taylor series expansion about the initial estimate (let us consider only functions g that are differentiable with respect to $\boldsymbol{\theta}$), we can approximate $g(\boldsymbol{x}_i, \boldsymbol{\theta})$ by a linear function of $\boldsymbol{\theta}$. Thus we obtain an approximating linear regression model in which an ordinary linear least-squares estimate of $\boldsymbol{\theta}$ can be computed. This process of computing an updated estimate based on a previous estimate can be repeated until, hopefully, the sequence of estimates converges. See Weisberg (1985, Section 12.1), Myers (1990, Chapter 9), Bates and Watts (1988), and Seber and Wild (1989).

The least-squares procedure in the preceding paragraph is most justifiable if the errors are assumed to be independent and all to have the same variance. However, the data may more closely follow a model in which the errors have unequal variances. In the model $y_i = g(\boldsymbol{x}_i, \boldsymbol{\theta}) + e_i$, let us allow the variance of e_i to depend on the explanatory variables, on the parameters in $\boldsymbol{\theta}$, and on other parameters as well. A common model of this type is the one that assumes $\mathrm{Var}(e_i) = \sigma^2 g(\boldsymbol{x}_i, \boldsymbol{\theta})^\lambda$, where σ and λ are unknown parameters. One can estimate the parameters by an iterative procedure similar to the one above except that weighted, rather than ordinary, linear least squares is used at each iteration. See Carroll and Ruppert (1988, Chapter 2) and Seber and Wild (1989, Section 2.8).

Generalized Linear Models. In a *generalized linear model* the regression function has the form $\mu_i = h(\boldsymbol{\beta}'\boldsymbol{x}_i)$, where h is a known function. The function h can be any strictly monotone (that is, either strictly increasing or strictly decreasing) differentiable function. Unless h is a linear function, this is a nonlinear regression function, but note that the explanatory variables and the parameters are interrelated in a linear way. The generalized linear model further assumes that the random variable y_i has a distribution with a probability density function (or probability mass function if y_i is discrete) of the form $f(y_i; \theta_i, \phi) = c(y_i, \phi)\exp\{a(\phi)[y_i\theta_i - b(\theta_i)]\}$. This includes normal, gamma, binomial, and Poisson distributions. The parameter vector $\boldsymbol{\beta}$ is linked to the distribution of y_i by the fact that $\mu_i = b'(\theta_i)$. To estimate the parameters, the maximum likelihood method is used. The linear form of the interrelationship between $\boldsymbol{x}_i$ and $\boldsymbol{\beta}$ and the exponential form of $f(y_i; \theta_i, \phi)$ allow some useful aspects of linear regression analysis to be carried over to the analysis of these more general models. See McCullagh and Nelder (1989) and Myers (1990, Section 7.6).

Nonparametric Smoothing. In nonlinear regression it is assumed that the form of the regression function is known. That is, the function g in $\mu_i = g(\boldsymbol{x}_i, \boldsymbol{\theta})$ is assumed known. A more general regression model is obtained by letting the form of the regression function be unknown. Such a model is especially appropriate for exploratory data analysis. Let us suppose $\mu_i = g(\boldsymbol{x}_i)$, where g is an unknown function. Estimating the function g can be viewed as "smoothing" a plot of the data. One tries to find a "smooth" function $\hat{g}(\boldsymbol{x})$ such that, in the $(p + 1)$-dimensional plot of the data, the points $(\boldsymbol{x}_i, \hat{g}(\boldsymbol{x}_i))$ are close to the data points $(\boldsymbol{x}_i, y_i)$. For example, one might take $\hat{g}(\boldsymbol{x})$ to be some sort of "local average", such as the average of all values

y_i such that $\boldsymbol{x}_i$ is within a certain neighborhood of $\boldsymbol{x}$. See Eubank (1988) and Härdle (1990). Such regression procedures are called *nonparametric* because no parametric model is assumed for the regression function, but note that they are unrelated to the nonparametric regression procedure described in Chapter 6.

Estimation by local averaging does not perform as well in multiple regression as it does in simple regression. This problem led to the development of *projection pursuit regression*, which estimates a multiple regression function through an iterative sequence of simple regression smoothings; see Friedman and Stuetzle (1981). The same idea has been used to generalize generalized linear models to *generalized additive models*. A generalized linear model assumes that $\mu_i = h(\boldsymbol{\beta}'\boldsymbol{x}_i) = h(\beta_1 x_{i1} + \cdots + \beta_p x_{ip})$, whereas a generalized additive model assumes only that $\mu_i = h(g_1(x_{i1}) + \cdots + g_p(x_{ip}))$, where $g_1, \ldots, g_p$ are unknown smooth functions; see Hastie and Tibshirani (1990).

Measurement Error in the Explanatory Variables. Another situation in which the linear regression model (1.1) may be inadequate is when the explanatory variables are subject to measurement error. Suppose that (1.1) is valid but that we are not able to observe the exact values of the explanatory variables X_j. Instead, due to measurement error, we observe $(*)\ Z_j = X_j + d_j$, where d_j is a random error. Unless the measurement errors are very small, it is not safe to proceed as if $Z_j = X_j$. If we do, the estimates of the regression coefficients will be biased. In simple regression, the estimate of the slope will be biased toward 0. See Draper and Smith (1981, Section 2.14) and Myers (1990, Section 7.8). Better estimates can be derived from the model defined by (1.1) and $(*)$ together; see Fuller (1987, Section 2.2).

REFERENCES

Bates, D. M., and D. G. Watts (1988). *Nonlinear Regression Analysis and Its Applications*. Wiley, New York.

Carroll, R. J., and D. Ruppert (1988). *Transformation and Weighting in Regression*. Chapman and Hall, New York.

Dodge, Y. (1984). Robust estimation of regression coefficients by minimizing a convex combination of least squares and least absolute deviations. *Computational Statistics Quarterly*, vol. 1, pp. 139–153.

Draper, N. R., and H. Smith (1981). *Applied Regression Analysis*, 2nd ed. Wiley, New York.

Eubank, R. L. (1988). *Smoothing Splines and Nonparametric Regression*. Marcel Dekker, New York.

Friedman, J. H., and W. Stuetzle (1981). Projection pursuit regression. *Journal of the American Statistical Association*, vol. 76, pp. 817–823.

Fuller, W. A. (1987). *Measurement Error Models*. Wiley, New York.

Gonin, R., and A. H. Money (1989). *Nonlinear L_p-Norm Estimation*. Marcel Dekker, New York.

Gunst, R. F., and R. L. Mason (1980). *Regression Analysis and Its Application: A Data-Oriented Approach*. Marcel Dekker, New York.

Hampel, F. R., E. M. Ronchetti, P. J. Rousseeuw, and W. A. Stahel (1986). *Robust Statistics: The Approach Based on Influence Functions*. Wiley, New York.

Härdle, W. (1990). *Applied Nonparametric Regression*. Cambridge University Press, Cambridge.

Hastie, T. J., and R. J. Tibshirani (1990). *Generalized Additive Models*. Chapman and Hall, London.

Hettmansperger, T. P. (1984). *Statistical Inference Based on Ranks*. Wiley, New York.

Huber, P. J. (1981). *Robust Statistics*. Wiley, New York.

Jennrich, R. I., and S. D. Oman (1986). How much does Stein estimation help in multiple regression? *Technometrics*, vol. 28, pp. 113–121.

Jurečková, J., and S. Portnoy (1987). Asymptotics for one-step M estimators in regression with applications to combining efficiency and high breakdown point. *Communications in Statistics A*, vol. 16, pp. 2187–2200.

Koenker, R., and G. Basset (1978). Regression quantiles. *Econometrica*, vol. 46, pp. 33–50.

Koenker, R., and S. Portnoy (1987). L-estimation for linear models. *Journal of the American Statistical Association*, vol. 82, pp. 851–857.

Lee, Y., and D. Birkes (1993). Shrinking toward submodels in regression. *Journal of Statistical Planning and Inference* (submitted).

McCullagh, P., and J. A. Nelder (1989). *Generalized Linear Models*, 2nd ed. Chapman and Hall, London.

McKean, J. W., and G. L. Sievers (1989). Rank scores suitable for analysis of linear models under asymmetric error distributions. *Technometrics*, vol. 31, pp. 207–218.

Myers, R. H. (1990). *Classical and Modern Regression with Applications*. PWS-Kent, Boston.

Oman, S. D. (1982). Shrinking towards subspaces in multiple linear regression. *Technometrics*, vol. 24, pp. 307–311.

Rolph, J. E. (1976). Choosing shrinkage estimators for regression problems. *Communications in Statistics A*, vol. 5, pp. 789–802.

Rousseeuw, P. J., and A. M. Leroy (1987). *Robust Regression and Outlier Detection*. Wiley, New York.

Seber, G. A. F. (1977). *Linear Regression Analysis*. Wiley, New York.

Seber, G. A. F., and C. J. Wild (1989). *Nonlinear Regression*. Wiley, New York.

Weisberg, S. (1985). *Applied Linear Regression*, 2nd ed. Wiley, New York.

Yohai, V. J., and R. H. Zamar (1988). High breakdown-point estimates of regression by means of the minimization of an efficient scale. *Journal of the American Statistical Association*, vol. 83, pp. 406–413.

APPENDIX

STUDENT'S t-DISTRIBUTION

The table contains two-tailed values t_{tab} such that Prob$[|t| \geq t_{tab}] = \alpha$ where t is a random variable having a t distribution with ν degrees of freedom.

ν	0.2	0.1	0.05	0.01	0.001
1	3.078	6.314	12.706	63.657	636.62
2	1.886	2.920	4.303	9.925	31.598
3	1.638	2.353	3.182	5.841	12.924
4	1.533	2.132	2.776	4.604	8.610
5	1.476	2.015	2.571	4.032	6.869
6	1.440	1.943	2.447	3.707	5.959
7	1.415	1.895	2.365	3.499	5.408
8	1.397	1.860	2.306	3.355	5.041
9	1.383	1.833	2.262	3.250	4.781
10	1.372	1.812	2.228	3.169	4.587
11	1.363	1.796	2.201	3.106	4.437
12	1.356	1.782	2.179	3.055	4.318
13	1.350	1.771	2.160	3.012	4.221
14	1.345	1.761	2.145	2.977	4.140
15	1.341	1.753	2.131	2.947	4.073
16	1.337	1.746	2.120	2.921	4.016
17	1.333	1.740	2.110	2.898	3.965
18	1.330	1.734	2.101	2.878	3.922
19	1.328	1.729	2.093	2.861	3.883
20	1.325	1.725	2.086	2.845	3.850
21	1.323	1.721	2.080	2.831	3.819
22	1.321	1.717	2.074	2.819	3.792
23	1.319	1.714	2.069	2.807	3.767
24	1.318	1.711	2.064	2.797	3.745
25	1.316	1.708	2.060	2.787	3.725
26	1.315	1.706	2.056	2.779	3.707
27	1.314	1.703	2.052	2.771	3.690
28	1.313	1.701	2.048	2.763	3.674
29	1.311	1.699	2.045	2.756	3.659
30	1.310	1.697	2.042	2.750	3.646
40	1.303	1.684	2.021	2.704	3.551
60	1.296	1.671	2.000	2.660	3.460
120	1.289	1.658	1.980	2.617	3.373
∞	1.282	1.645	1.960	2.576	3.291

Reproduced with permission from E. S. Pearson and H. O. Hartley (1966), *Biometrika Tables for Statisticians*, Vol. 1, 3rd ed., London: Cambridge University.

F-DISTRIBUTION

The table contains values F_{tab} such that $\text{Prob}[F \geq F_{tab}] = \alpha$ where F is a random variable having an F distribution with ν_1 and ν_2 degrees of freedom.

$\alpha = 0.05$

Degrees of freedom for numerator

Degrees of Freedom for Denominator ν_2 \ ν_1	1	2	3	4	5	6	7	8	9
1	161.4	199.5	215.7	224.6	230.2	234.0	236.8	238.9	240.5
2	18.51	19.00	19.16	19.25	19.30	19.33	19.35	19.37	19.38
3	10.13	9.55	9.28	9.12	9.01	8.94	8.89	8.85	8.81
4	7.71	6.94	6.59	6.39	6.26	6.16	6.09	6.04	6.00
5	6.61	5.79	5.41	5.19	5.05	4.95	4.88	4.82	4.77
6	5.99	5.14	4.76	4.53	4.39	4.28	4.21	4.15	4.10
7	5.59	4.74	4.35	4.12	3.97	3.87	3.79	3.73	3.68
8	5.32	4.46	4.07	3.84	3.69	3.58	3.50	3.44	3.39
9	5.12	4.26	3.86	3.63	3.48	3.37	3.29	3.23	3.18
10	4.96	4.10	3.71	3.48	3.33	3.22	3.14	3.07	3.02
11	4.84	3.98	3.59	3.36	3.20	3.09	3.01	2.95	2.90
12	4.75	3.89	3.49	3.26	3.11	3.00	2.91	2.85	2.80
13	4.67	3.81	3.41	3.18	3.03	2.92	2.83	2.77	2.71
14	4.60	3.74	3.34	3.11	2.96	2.85	2.76	2.70	2.65
15	4.54	3.68	3.29	3.06	2.90	2.79	2.71	2.64	2.59
16	4.49	3.63	3.24	3.01	2.85	2.74	2.66	2.59	2.54
17	4.45	3.59	3.20	2.96	2.81	2.70	2.61	2.55	2.49
18	4.41	3.55	3.16	2.93	2.77	2.66	2.58	2.51	2.46
19	4.38	3.52	3.13	2.90	2.74	2.63	2.54	2.48	2.42
20	4.35	3.49	3.10	2.87	2.71	2.60	2.51	2.45	2.39
21	4.32	3.47	3.07	2.84	2.68	2.57	2.49	2.42	2.37
22	4.30	3.44	3.05	2.82	2.66	2.55	2.46	2.40	2.34
23	4.28	3.42	3.03	2.80	2.64	2.53	2.44	2.37	2.32
24	4.26	3.40	3.01	2.78	2.62	2.51	2.42	2.36	2.30
25	4.24	3.39	2.99	2.76	2.60	2.49	2.40	2.34	2.28
26	4.23	3.37	2.98	2.74	2.59	2.47	2.39	2.32	2.27
27	4.21	3.35	2.96	2.73	2.57	2.46	2.37	2.31	2.25
28	4.20	3.34	2.95	2.71	2.56	2.45	2.36	2.29	2.24
29	4.18	3.33	2.93	2.70	2.55	2.43	2.35	2.28	2.22
30	4.17	3.32	2.92	2.69	2.53	2.42	2.33	2.27	2.21
40	4.08	3.23	2.84	2.61	2.45	2.34	2.25	2.18	2.12
60	4.00	3.15	2.76	2.53	2.37	2.25	2.17	2.10	2.04
120	3.92	3.07	2.68	2.45	2.29	2.17	2.09	2.02	1.96
∞	3.84	3.00	2.60	2.37	2.21	2.10	2.01	1.94	1.88

10	12	15	20	24	30	40	60	120	∞
241.9	243.9	245.9	248.0	249.1	250.1	251.1	252.2	253.3	254.3
19.40	19.41	19.43	19.45	19.45	19.46	19.47	19.48	19.49	19.50
8.79	8.74	8.70	8.66	8.64	8.62	8.59	8.57	8.55	8.53
5.96	5.91	5.86	5.80	5.77	5.75	5.72	5.69	5.66	5.63
4.74	4.68	4.62	4.56	4.53	4.50	4.46	4.43	4.40	4.36
4.06	4.00	3.94	3.87	3.84	3.81	3.77	3.74	3.70	3.67
3.64	3.57	3.51	3.44	3.41	3.38	3.34	3.30	3.27	3.23
3.35	3.28	3.22	3.15	3.12	3.08	3.04	3.01	2.97	2.93
3.14	3.07	3.01	2.94	2.90	2.86	2.83	2.79	2.75	2.71
2.98	2.91	2.85	2.77	2.74	2.70	2.66	2.62	2.58	2.54
2.85	2.79	2.72	2.65	2.61	2.57	2.53	2.49	2.45	2.40
2.75	2.69	2.62	2.54	2.51	2.47	2.43	2.38	2.34	2.30
2.67	2.60	2.53	2.46	2.42	2.38	2.34	2.30	2.25	2.21
2.60	2.53	2.46	2.39	2.35	2.31	2.27	2.22	2.18	2.13
2.54	2.48	2.40	2.33	2.29	2.25	2.20	2.16	2.11	2.07
2.49	2.42	2.35	2.28	2.24	2.19	2.15	2.11	2.06	2.01
2.45	2.38	2.31	2.23	2.19	2.15	2.10	2.06	2.01	1.96
2.41	2.34	2.27	2.19	2.15	2.11	2.06	2.02	1.97	1.92
2.38	2.31	2.23	2.16	2.11	2.07	2.03	1.98	1.93	1.88
2.35	2.28	2.20	2.12	2.08	2.04	1.99	1.95	1.90	1.84
2.32	2.25	2.18	2.10	2.05	2.01	1.96	1.92	1.87	1.81
2.30	2.23	2.15	2.07	2.03	1.98	1.94	1.89	1.84	1.78
2.27	2.20	2.13	2.05	2.01	1.96	1.91	1.86	1.81	1.76
2.25	2.18	2.11	2.03	1.98	1.94	1.89	1.84	1.79	1.73
2.24	2.16	2.09	2.01	1.96	1.92	1.87	1.82	1.77	1.71
2.22	2.15	2.07	1.99	1.95	1.90	1.85	1.80	1.75	1.69
2.20	2.13	2.06	1.97	1.93	1.88	1.84	1.79	1.73	1.67
2.19	2.12	2.04	1.96	1.91	1.87	1.82	1.77	1.71	1.65
2.18	2.10	2.03	1.94	1.90	1.85	1.81	1.75	1.70	1.64
2.16	2.09	2.01	1.93	1.89	1.84	1.79	1.74	1.68	1.62
2.08	2.00	1.92	1.84	1.79	1.74	1.69	1.64	1.58	1.51
1.99	1.92	1.84	1.75	1.70	1.65	1.59	1.53	1.47	1.39
1.91	1.83	1.75	1.66	1.61	1.55	1.50	1.43	1.35	1.25
1.83	1.75	1.67	1.57	1.52	1.46	1.39	1.32	1.22	1.00

F-DISTRIBUTION

The table contains values F_{tab} such that $\text{Prob}[F \geq F_{tab}] = \alpha$ where F is a random variable having an F distribution with ν_1 and ν_2 degrees of freedom.

$\alpha = 0.01$

Degrees of freedom for numerator

Degrees of Freedom for Denominator ν_2 \ ν_1	1	2	3	4	5	6	7	8	9
1	4052	4999.5	5403	5625	5764	5859	5928	5982	6022
2	98.50	99.00	99.17	99.25	99.30	99.33	99.36	99.37	99.39
3	34.12	30.82	29.46	28.71	28.24	27.91	27.67	27.49	27.35
4	21.20	18.00	16.69	15.98	15.52	15.21	14.98	14.80	14.66
5	16.26	13.27	12.06	11.39	10.97	10.67	10.46	10.29	10.16
6	13.75	10.92	9.78	9.15	8.75	8.47	8.26	8.10	7.98
7	12.25	9.55	8.45	7.85	7.46	7.19	6.99	6.84	6.72
8	11.26	8.65	7.59	7.01	6.63	6.37	6.18	6.03	5.91
9	10.56	8.02	6.99	6.42	6.06	5.80	5.61	5.47	5.35
10	10.04	7.56	6.55	5.99	5.64	5.39	5.20	5.06	4.94
11	9.65	7.21	6.22	5.67	5.32	5.07	4.89	4.74	4.63
12	9.33	6.93	5.95	5.41	5.06	4.82	4.64	4.50	4.39
13	9.07	6.70	5.74	5.21	4.86	4.62	4.44	4.30	4.19
14	8.86	6.51	5.56	5.04	4.69	4.46	4.28	4.14	4.03
15	8.68	6.36	5.42	4.89	4.56	4.32	4.14	4.00	3.89
16	8.53	6.23	5.29	4.77	4.44	4.20	4.03	3.89	3.78
17	8.40	6.11	5.18	4.67	4.34	4.10	3.93	3.79	3.68
18	8.29	6.01	5.09	4.58	4.25	4.01	3.84	3.71	3.60
19	8.18	5.93	5.01	4.50	4.17	3.94	3.77	3.63	3.52
20	8.10	5.85	4.94	4.43	4.10	3.87	3.70	3.56	3.46
21	8.02	5.78	4.87	4.37	4.04	3.81	3.64	3.51	3.40
22	7.95	5.72	4.82	4.31	3.99	3.76	3.59	3.45	3.35
23	7.88	5.66	4.76	4.26	3.94	3.71	3.54	3.41	3.30
24	7.82	5.61	4.72	4.22	3.90	3.67	3.50	3.36	3.26
25	7.77	5.57	4.68	4.18	3.85	3.63	3.46	3.32	3.22
26	7.72	5.53	4.64	4.14	3.82	3.59	3.42	3.29	3.18
27	7.68	5.49	4.60	4.11	3.78	3.56	3.39	3.26	3.15
28	7.64	5.45	4.57	4.07	3.75	3.53	3.36	3.23	3.12
29	7.60	5.42	4.54	4.04	3.73	3.50	3.33	3.20	3.09
30	7.56	5.39	4.51	4.02	3.70	3.47	3.30	3.17	3.07
40	7.31	5.18	4.31	3.83	3.51	3.29	3.12	2.99	2.89
60	7.08	4.98	4.13	3.65	3.34	3.12	2.95	2.82	2.72
120	6.85	4.79	3.95	3.48	3.17	2.96	2.79	2.66	2.56
∞	6.63	4.61	3.78	3.32	3.02	2.80	2.64	2.51	2.41

10	12	15	20	24	30	40	60	120	∞
6056	6106	6157	6209	6235	6261	6287	6313	6339	6366
99.40	99.42	99.43	99.45	99.46	99.47	99.47	99.48	99.49	99.50
27.23	27.05	26.87	26.69	26.60	26.50	26.41	26.32	26.22	26.13
14.55	14.37	14.20	14.02	13.93	13.84	13.75	13.65	13.56	13.46
10.05	9.89	9.72	9.55	9.47	9.38	9.29	9.20	9.11	9.02
8.87	7.72	7.56	7.40	7.31	7.23	7.14	7.06	6.97	6.88
6.62	6.47	6.31	6.16	6.07	5.99	5.91	5.82	5.74	5.65
5.81	5.67	5.52	5.36	5.28	5.20	5.12	5.03	4.95	4.86
5.26	5.11	4.96	4.81	4.73	4.65	4.57	4.48	4.40	4.31
4.85	4.71	4.56	4.41	4.33	4.25	4.17	4.08	4.00	3.91
4.54	4.40	4.25	4.10	4.02	3.94	3.86	2.78	3.69	3.60
4.30	4.16	4.01	3.86	3.78	3.70	3.62	3.54	3.45	3.36
4.10	3.96	3.82	3.66	3.59	3.51	3.43	3.34	3.25	3.17
3.94	3.80	3.66	3.51	3.43	3.35	3.27	3.18	3.09	3.00
3.80	3.67	3.52	3.37	3.29	3.21	3.13	3.05	2.96	2.87
3.69	3.55	3.41	3.26	3.18	3.10	3.02	2.93	2.84	2.75
3.59	3.46	3.31	3.16	3.08	3.00	2.92	2.83	2.75	2.65
3.51	3.37	3.23	3.08	3.00	2.92	2.84	2.75	2.66	2.57
3.43	3.30	3.15	3.00	2.92	2.84	2.76	2.67	2.58	2.49
3.37	3.23	3.09	2.94	2.86	2.78	2.69	2.61	2.52	2.42
3.31	3.17	3.03	2.88	2.80	2.72	2.64	2.55	2.46	2.36
3.26	3.12	2.98	2.83	2.75	2.67	2.58	2.50	2.40	2.31
3.21	3.07	2.93	2.78	2.70	2.62	2.54	2.45	2.35	2.26
3.17	3.03	2.89	2.74	2.66	2.58	2.49	2.40	2.31	2.21
3.13	2.99	2.85	2.70	2.62	2.54	2.45	2.36	2.27	2.17
3.09	2.96	2.81	2.66	2.58	2.50	2.42	2.33	2.23	2.13
3.06	2.93	2.78	2.63	2.55	2.47	2.38	2.29	2.20	2.10
3.03	2.90	2.75	2.60	2.52	2.44	2.35	2.26	2.17	2.06
3.00	2.87	2.73	2.57	2.49	2.41	2.33	2.23	2.14	2.03
2.98	2.84	2.70	2.55	2.47	2.39	2.30	2.21	2.11	2.01
2.80	2.66	2.52	2.37	2.29	2.20	2.11	2.02	1.92	1.80
2.63	2.50	2.35	2.20	2.12	2.03	1.94	1.84	1.73	1.60
2.47	2.34	2.19	2.03	1.95	1.86	1.76	1.66	1.53	1.38
2.32	2.18	2.04	1.88	1.79	1.70	1.59	1.47	1.32	1.00

F-DISTRIBUTION

The table contains values F_{tab} such that $\text{Prob}[F \geq F_{tab}] = \alpha$ where F is a random variable having an F distribution with ν_1 and ν_2 degrees of freedom.

$\alpha = 0.001$

Degrees of freedom for numerator

Degrees of Freedom for Denominator

ν_2 \ ν_1	1	2	3	4	5	6	7	8	9
1	4053*	5000*	5404*	5625*	5764*	5859*	5929*	5981*	6023*
2	998.5	999.0	999.2	999.2	999.3	999.3	999.4	999.4	999.4
3	167.0	148.5	141.1	137.1	134.6	132.8	131.6	130.6	129.9
4	74.14	61.25	56.18	53.44	51.71	50.53	49.66	49.00	48.47
5	47.18	37.12	33.20	31.09	29.75	28.84	28.16	27.64	27.24
6	35.51	27.00	23.70	21.92	20.81	20.03	19.46	19.03	18.69
7	29.25	21.69	18.77	17.19	16.21	15.52	15.02	14.63	14.33
8	25.42	18.49	15.83	14.39	13.49	12.86	12.40	12.04	11.77
9	22.86	16.39	13.90	12.56	11.71	11.13	10.70	10.37	10.11
10	21.04	14.91	12.55	11.28	10.48	9.92	9.52	9.20	8.96
11	19.69	13.81	11.56	10.35	9.58	9.05	8.66	8.35	8.12
12	18.64	12.97	10.80	9.63	8.89	8.38	8.00	7.71	7.48
13	17.81	12.31	10.21	9.07	8.35	7.86	7.49	7.21	6.98
14	17.14	11.78	9.73	8.62	7.92	7.43	7.08	6.80	6.58
15	16.59	11.34	9.34	8.25	7.57	7.09	6.74	6.47	6.26
16	16.12	10.97	9.00	7.94	7.27	6.81	6.46	6.19	5.98
17	15.72	10.66	8.73	7.68	7.02	6.56	6.22	5.96	5.75
18	15.38	10.39	8.49	7.46	6.81	6.35	6.02	5.76	5.56
19	15.08	10.16	8.28	7.26	6.62	6.18	5.85	5.59	5.39
20	14.82	9.95	8.10	7.10	6.46	6.02	5.69	5.44	5.24
21	14.59	9.77	7.94	6.95	6.32	5.88	5.56	5.31	5.11
22	14.38	9.61	7.80	6.81	6.19	5.76	5.44	5.19	4.99
23	14.19	9.47	7.67	6.69	6.08	5.65	5.33	5.09	4.89
24	14.03	9.34	7.55	6.59	5.98	5.55	5.23	4.99	4.80
25	13.88	9.22	7.45	6.49	5.88	5.46	5.15	4.91	4.71
26	13.74	9.12	7.36	6.41	5.80	5.38	5.07	4.83	4.64
27	13.61	9.02	7.27	6.33	5.73	5.31	5.00	4.76	4.57
28	13.50	8.93	7.19	6.25	5.66	5.24	4.93	4.69	4.50
29	13.39	8.85	7.12	6.19	5.59	5.18	4.87	4.64	4.45
30	13.29	8.77	7.05	6.12	5.53	5.12	4.82	4.58	4.39
40	12.61	8.25	6.60	5.70	5.13	4.73	4.44	4.21	4.02
60	11.97	7.76	6.17	5.31	4.76	4.37	4.09	3.87	3.69
120	11.38	7.32	5.79	4.95	4.42	4.04	3.77	3.55	3.38
∞	10.83	6.91	5.42	4.62	4.10	3.74	3.47	3.27	3.10

Reproduced with permission from E. S. Pearson and H. O. Hartley (1966), *Biometrika Tables for Statisticians*, Vol. 1, 3rd ed., London: Cambridge University.

*Multiply these numbers by 100.

10	12	15	20	24	30	40	60	120	∞
6056*	6107*	6158*	6209*	6235*	6261*	6287*	6313*	6340*	6366*
999.4	999.4	999.4	999.4	999.5	999.5	999.5	999.5	999.5	999.5
129.2	128.3	127.4	126.4	125.9	125.4	125.0	124.5	124.0	123.5
48.05	47.41	46.76	46.10	45.77	45.43	45.09	44.75	44.40	44.05
26.92	26.42	25.91	25.39	25.14	24.87	24.60	24.33	24.06	23.79
18.41	17.99	17.56	17.12	16.89	16.67	16.44	16.21	15.99	15.75
14.08	13.71	13.32	12.93	12.73	12.53	12.33	12.12	11.91	11.70
11.54	11.19	10.84	10.48	10.30	10.11	9.92	9.73	9.53	9.33
9.89	9.57	9.24	8.90	8.72	8.55	8.37	8.19	8.00	7.81
8.75	8.45	8.13	7.80	7.64	7.47	7.30	7.12	6.94	6.76
7.92	7.63	7.32	7.01	6.85	6.68	6.52	6.35	6.17	6.00
7.29	7.00	6.71	6.40	6.25	6.09	5.93	5.76	5.59	5.42
6.80	6.52	6.23	5.93	5.78	5.63	5.47	5.30	5.14	4.97
6.40	6.13	5.85	5.56	5.41	5.25	5.10	4.94	4.77	4.60
6.08	5.81	5.54	5.25	5.10	4.95	4.80	4.64	4.47	4.31
5.81	5.55	5.27	4.99	4.85	4.70	4.54	4.39	4.23	4.06
5.58	5.32	5.05	4.78	4.63	4.48	4.33	4.18	4.02	3.85
5.39	5.13	4.87	4.59	4.45	4.30	4.15	4.00	3.84	3.67
5.22	4.97	4.70	4.43	4.29	4.14	3.99	3.84	3.68	3.51
5.08	4.82	4.56	4.29	4.15	4.00	3.86	3.70	3.54	3.38
4.95	4.70	4.44	4.17	4.03	3.88	3.74	3.58	3.42	3.26
4.83	4.58	4.33	4.06	3.92	3.78	3.63	3.48	3.32	3.15
4.73	4.48	4.23	3.96	3.82	3.68	3.53	3.38	3.22	3.05
4.64	4.39	4.14	3.87	3.74	3.59	3.45	3.29	3.14	2.97
4.56	4.31	4.06	3.79	3.66	3.52	3.37	3.22	3.06	2.89
4.48	4.24	3.99	3.72	3.59	3.44	3.30	3.15	2.99	2.82
4.41	4.17	3.92	3.66	3.52	3.38	3.23	3.08	2.92	2.75
4.35	4.11	3.86	3.60	3.46	3.32	3.18	3.02	2.86	2.69
4.29	4.05	3.80	3.54	3.41	3.27	3.12	2.97	2.81	2.64
4.24	4.00	3.75	3.49	3.36	3.22	3.07	2.92	2.76	2.59
3.87	3.64	3.40	3.15	3.01	2.87	2.73	2.57	2.41	2.23
3.54	3.31	3.08	2.83	2.69	2.55	2.41	2.25	2.08	1.89
3.24	3.02	2.78	2.53	2.40	2.26	2.11	1.95	1.76	1.54
2.96	2.74	2.51	2.27	2.13	1.99	1.84	1.66	1.45	1.00

CHI-SQUARED DISTRIBUTION

The table shows percentage points of the chi-squared distribution; tabled values are $\chi^2(\alpha; n)$ such that

$$\text{prob}\{\chi^2(n) \text{ variate} \geq \chi^2(\alpha; n)\} = \alpha.$$

The table entries were computed using subroutine MDCHI from the IMSL (1977) library at the University of Minnesota.

	α				
d.f.	0.20	0.10	0.05	0.01	0.001
1	1.64	2.71	3.84	6.64	10.81
2	3.22	4.60	5.99	9.22	13.69
3	4.64	6.25	7.82	11.32	16.29
4	5.99	7.78	9.49	13.28	18.43
5	7.29	9.24	11.07	15.09	20.75
6	8.56	10.65	12.60	16.81	22.68
7	9.80	12.02	14.07	18.47	24.53
8	11.03	13.36	15.51	20.08	26.32
9	12.24	14.69	16.93	21.65	28.06
10	13.44	15.99	18.31	23.19	29.76
11	14.63	17.28	19.68	24.75	31.43
12	15.81	18.55	21.03	26.25	33.07
13	16.99	19.81	22.37	27.72	34.68
14	18.15	21.07	23.69	29.17	36.27
15	19.31	22.31	25.00	30.61	37.84
16	20.47	23.55	26.30	32.03	39.39
17	21.62	24.77	27.59	33.44	40.93
18	22.76	25.99	28.88	34.83	42.44
19	23.90	27.21	30.15	36.22	43.95
20	25.04	28.42	31.42	37.59	45.44
21	26.17	29.62	32.68	38.96	46.92
22	27.30	30.82	33.93	40.31	48.39
23	28.43	32.01	35.18	41.66	49.85
24	29.56	33.20	36.42	43.00	51.29
25	30.68	34.38	37.66	44.34	52.73
26	31.80	35.57	38.89	45.66	54.16
27	32.91	36.74	40.12	46.99	55.58
28	34.03	37.92	41.34	48.30	57.00
29	35.14	39.09	42.56	49.61	58.41
30	36.25	40.26	43.78	50.91	59.81
40	47.26	51.80	55.75	63.71	73.49
50	58.16	63.16	67.50	76.17	86.74
60	68.97	74.39	79.08	88.40	99.68
70	79.71	85.52	90.53	100.44	112.38
80	90.40	96.57	101.88	112.34	124.90
90	101.05	107.56	113.14	124.13	137.27
100	111.66	118.49	124.34	135.82	149.50

Index

Applied Probability and Statistics (Continued)

JOHNSON and KOTZ · Distributions in Statistics
Discrete Distributions
Continuous Univariate Distributions—1
Continuous Univariate Distributions—2
Continuous Multivariate Distributions

JUDGE, GRIFFITHS, HILL, LÜTKEPOHL, and LEE · The Theory and Practice of Econometrics, *Second Edition*

JUDGE, HILL, GRIFFITHS, LÜTKEPOHL, and LEE · Introduction to the Theory and Practice of Econometrics, *Second Edition*

KALBFLEISCH and PRENTICE · The Statistical Analysis of Failure Time Data

KASPRZYK, DUNCAN, KALTON, and SINGH · Panel Surveys

KISH · Statistical Design for Research

KISH · Survey Sampling

LAWLESS · Statistical Models and Methods for Lifetime Data

LEBART, MORINEAU, and WARWICK · Multivariate Descriptive Statistical Analysis: Correspondence Analysis and Related Techniques for Large Matrices

LEE · Statistical Methods for Survival Data Analysis, *Second Edition*

LePAGE and BILLARD · Exploring the Limits of Bootstrap

LEVY and LEMESHOW · Sampling of Populations: Methods and Applications

LINHART and ZUCCHINI · Model Selection

LITTLE and RUBIN · Statistical Analysis with Missing Data

MAGNUS and NEUDECKER · Matrix Differential Calculus with Applications in Statistics and Econometrics

MAINDONALD · Statistical Computation

MALLOWS · Design, Data, and Analysis by Some Friends of Cuthbert Daniel

MANN, SCHAFER, and SINGPURWALLA · Methods for Statistical Analysis of Reliability and Life Data

MASON, GUNST, and HESS · Statistical Design and Analysis of Experiments with Applications to Engineering and Science

McLACHLAN · Discriminant Analysis and Statistical Pattern Recognition

MILLER · Survival Analysis

MONTGOMERY and PECK · Introduction to Linear Regression Analysis, *Second Edition*

NELSON · Accelerated Testing, Statistical Models, Test Plans, and Data Analyses

NELSON · Applied Life Data Analysis

OCHI · Applied Probability and Stochastic Processes in Engineering and Physical Sciences

OKABE, BOOTS, and SUGIHARA · Spatial Tesselations: Concepts and Applications of Voronoi Diagrams

OSBORNE · Finite Algorithms in Optimization and Data Analysis

PANKRATZ · Forecasting with Dynamic Regression Models

PANKRATZ · Forecasting with Univariate Box-Jenkins Models: Concepts and Cases

RACHEV · Probability Metrics and the Stability of Stochastic Models

RÉNYI · A Diary on Information Theory

RIPLEY · Spatial Statistics

RIPLEY · Stochastic Simulation

ROSS · Introduction to Probability and Statistics for Engineers and Scientists

ROUSSEEUW and LEROY · Robust Regression and Outlier Detection

RUBIN · Multiple Imputation for Nonresponse in Surveys

RYAN · Statistical Methods for Quality Improvement

SCHUSS · Theory and Applications of Stochastic Differential Equations

SCOTT · Multivariate Density Estimation: Theory, Practice, and Visualization

SEARLE · Linear Models

SEARLE · Linear Models for Unbalanced Data

SEARLE · Matrix Algebra Useful for Statistics

SEARLE, CASELLA, and McCULLOCH · Variance Components

SKINNER, HOLT, and SMITH · Analysis of Complex Surveys

*Now available in a lower priced paperback edition in the Wiley Classics Library.